Electronic Diesel Engine Controls

SP-781

NEW ENGLAND INSTITUTE
OF TECHNOLOGY
LEARNING RESOURCES CENTER

The papers included in this volume are abstracted and indexed in the SAE Global Mobility Database.

Published by:
Society of Automotive Engineers, Inc.
400 Commonwealth Drive
Warrendale, PA 15096-0001
February 1989

Permission to photocopy for internal or personal use, or the internal or personal use of specific clients, is granted by SAE for libraries and other users registered with the Copyright Clearance Center (CCC), provided that the base fee of $3.00 per copy is paid directly to CCC, 21 Congress St., Salem, MA 01970. Special requests should be addressed to the SAE Publications Division. 0-89883-438-4/89 $3.00

No part of this publication may be reproduced in any form, in an electronic retrieval system or otherwise, without the prior written permission of the publisher.

ISBN 0-89883-438-4
SAE/SP-89/781
Library of Congress Catalog Card Number: 88-63814
Copyright 1989 Society of Automotive Engineers, Inc.

PREFACE

The continuing development of the application of modern electronics applied to engines is clearly illustrated by the papers in this session. The vital practical details of how to implement the current generation electronics are accompanied by other papers which look to the future in both hardware and software.

Modern electronics opens many windows of opportunity. The application of modern control methods is particularly important in assisting the development engineer in his task.

I am sure that any engineer with an interest in this area will find much of value in this book.

Bernard J. Challen
Session Organizer
Ricardo Consulting Engineers

TABLE OF CONTENTS

890387 **Design of Diesel Smoke Feedback Control Using a Combination of PI Control Algorithm and Performance Optimization** 1
    G. Hong and N. Collings
        Engineering Dept. Cambridge Univ.,
            England

890389 **Engine Modelling for Diesel Smoke Feedback Control System Design** 7
    G. Hong and N. Collings
        Engineering Dept. Cambridge Univ.,
            England

890391 **Electronic Controls for John Deere Diesel Engines** 13
    Marvin K. Farr
        Engine Engineering
            John Deere Engine Group

890392 **Control Design for a Differential Compound Engine** 23
    J. Hall and F.J. Wallace
        School of Mechanical Engineering
            University of Bath,
              United Kingdom

890393 **Transient Response and Route Simulations for Heavy Vehicles with Alternative Engine-Transmission Systems** 47
    Zou Dequan, M. Rezaian, and F.J. Wallace
        School of Mechanical Engineering
            University of Bath

890394 **Dynamic Simulation of a Turbocharged/Intercooled Diesel Engine with Rack-Actuated Electronic Fuel Control System** 67
    Charles I. Rackmil and Paul N. Blumberg
        Integral Technologies Inc.
            Westmont, Il.

890395 **Optimisation of Heavy-Duty Diesel Engine Transient Emissions by Advanced Control of a Variable Geometry Turbocharger** 79
    A.D. Pilley, A.D. Noble, A.J. Beaumont, J.R. Needham, and B.C. Porter
        Ricardo Consulting Engineers Ltd.

890396 **An Investigation of Cylinder Pressure as Feedback for Control of Internal Combustion Engines** 91
    Peter Wibberley and Christopher A. Clark
        Ricardo Consulting Engineers Ltd.

890397 **A Complete Engine Diagnostic System for Military Application** 97
    C. Operti
        IVECO
    W. Duss and J. W. Freestone
        DERECO

890387

Design of Diesel Smoke Feedback Control Using a Combination of PI Control Algorithm and Performance Optimization

G. Hong and N. Collings
Engineering Dept.
Cambridge Univ.,
England

ABSTRACT

A novel smoke sensor was used to realize smoke feedback control on a diesel engine. The controller design based on a combination of PI control algorithm and the engine performance optimization is described. Experimental results demonstrate how this control system behave to meet both of the speed and smoke requirements during engine transients.

ONE FOCUS OF RESEARCH on engine control has been to meet stringent requirements for fuel-efficient operation subject to constrains on emissions. A survey of engine control shows that some control algorithms, such as PID control[1], optimal control[2,3] and self-adaptive control[4,5] which has been developed recently, have been used in the simulation or practice of many engine control projects. However, further application of control theory to a multivariable engine system is still a fairly open problem.

Feedback control is usually deemed to be a better way than open-loop control to accomplish engine control when accurate set points are required. For some years the research on the sensors, the basic components of the feedback control system, has been the centre of engine control activities.

This paper describes a feedback control system for the control of diesel smoke based on a sensor which produces an electronic signal representing smoke level[6]. The objective of this research was to limit the smoke to a present minimum level during transients. Aimed at this objective, the engine system included two inputs (required engine speed and smoke) and two outputs (transient engine speed and smoke). Unlike a conventional design, it was designed with a combination of a preliminary analysis of the control problems and optimization of an appropriate interpretation of the overall performance specification. A PI algorithm was used to initially design the controllers, and then objective functions were chosen for optimization of the closed loop behaviour together with the determination of the controllers' coefficients. Experimental results are presented to demonstrate how this engine control system performs to approach the objective.

DESCRIPTION OF THE ENGINE CONTROL SYSTEM

The engine control system consisted of four components:
1) controlled engine: AEC diesel engine, DI, single-cylinder displacement;
2) computer: PDP11/23-PLUS with 20kHz ADC and DAC;
3) sensors: smoke sensor described in [6,7]; speed sensor: a tachogenerator;
4) actuator: a servo-motor acting on the fuel rack.

Two outputs of the control system were the engine speed and smoke. The input actuation was the fuel supply. The models necessary for the controller design were those of speed-to-fuel rack and smoke-to-fuel rack. The following are the transfer functions of these two models at engine condition of 800RPM speed and 30Nm torque, which are obtained from engine modelling work described in the accompanying paper[8] – the first stage of engine smoke control.

$$\begin{pmatrix} y_1(t) \\ y_2(t) \end{pmatrix} = \begin{pmatrix} 1.5375 & 0 \\ 0 & 1.0672 \end{pmatrix} \begin{pmatrix} y_1(t-1) \\ y_2(t-1) \end{pmatrix}$$
$$+ \begin{pmatrix} -0.5627 & 0 \\ 0 & -0.2084 \end{pmatrix} \begin{pmatrix} y_1(t-2) \\ y_2(t-2) \end{pmatrix}$$
$$+ \begin{pmatrix} 0.0145 \\ 0 \end{pmatrix} u(t-nd1) + \begin{pmatrix} 0 \\ 0.0451 \end{pmatrix} u(t-nd2)$$

where y_1 is the engine speed output, y_2 is the engine smoke output and u is the actuation which acts on the fuel rack. $nd1$ and $nd2$ define the time delay.

CONTROL STRATEGY

The diagram describing the control strategy is shown in Fig 1. $R_1(z^{-1})$ is the speed requirement, and $R_2(z^{-1})$ the smoke requirement. Two feedback outputs are $Y_1(z^{-1})$, the actual speed, and $Y_2(z^{-1})$, the actual smoke. $C_1(z^{-1})$ and $C_2(z^{-1})$ are the controllers to process the speed or smoke error signals. The speed requirement, $R_1(z^{-1})$, is a step input with positive amplitude. The smoke requirement, $R_2(z^{-1})$, is the smoke level at steady-state engine condition. It can be treated as a step input with zero amplitude.

As the smoke requirement, $R_2(z^{-1})$, is zero, the transfer functions between outputs and inputs are:

$$\begin{pmatrix} Y_1(z^{-1}) \\ Y_2(z^{-1}) \end{pmatrix} = \begin{pmatrix} Q_1(z^{-1}) \\ Q_2(z^{-1}) \end{pmatrix} R_1(z^{-1})$$

where

$$Q_1(z^{-1}) = \frac{C_1(z^{-1})G_1(z^{-1})}{1 + C_1(z^{-1})G_1(z^{-1}) + C_2(z^{-1})G_2(z^{-1})}$$

$$Q_2(z^{-1}) = \frac{C_1(z^{-1})G_2(z^{-1})}{1 + C_1(z^{-1})G_1(z^{-1}) + C_2(z^{-1})G_2(z^{-1})}$$

where $G_1(z^{-1})$ is the transfer function of speed-to-fuel rack and $G_2(z^{-1})$ smoke-to-fuel rack.

These two transfer functions have the same characteristic equations. This means that the stabilities of the two outputs correspond to the speed requirement correspond.

In this control system, the actual speed is compared with the required one while the actual smoke is compared with the requirement in a parallel way. Two error signals are processed in their respect controllers, then combined to give the command to the fuel rack.

CONTROLLER DESIGN

Computational simulation of the designed system

Simulation study was made on the models of the real engine systems in order to
(a) select the weights in the objective functions through observing the engine optimal process
(b) determine a set of coefficients for the controllers essential for optimizing the performance
(c) observe the behaviour of the designed control system in terms of the engine performance during transients and required measurement of the control system

It was found from the attempts to model the smoke-to-fuel rack transfer function that it was highly nonlinear. For approximating its nonlinearity, the smoke level was calculated using two groups of parameters of the transfer function. It is

$$y_2(t+1) = \begin{cases} a_{1(2)}y_2(t) + a_{2(2)}y_2(t-1) + b_{0(2)}u(t-T_{d2}) & \text{when } u \leq u_m \\ a'_{1(2)}y_2(t) + a'_{2(2)}y_2(t-1) + b'_{0(2)}u(t-T_{d2}) & \text{when } u \geq u_m \end{cases}$$

Two groups of parameters were obtained by modelling with different sizes of the inputs in our previous work.

Design of the PI controller for speed loop

Design was started with the consideration of the speed loop at first, then the smoke control based on the initially optimized speed loop was designed.

The controller 1, $C(z^{-1})$, in Fig. 1 was chosen as a PI controller. The model of speed-to-fuel rack is second order and can be represented as follows,

$$G_1(z^{-1}) = \frac{b_{0(1)}z}{(z - p_{1(1)})(z - p_{2(1)})}$$

here $p_{1(1)} = 0.9368$, $p_{2(1)} = 0.6006$.

The principle of the design can be described as follows,

The transfer function of controller 1 is,

$$C_1(z^{-1}) = k_{p1} + k_{i1}\frac{z}{z-1}$$

$$= (k_{p1} + k_{i1})\frac{z + \frac{k_{p1}}{k_{p1}+k_{i1}}}{z-1}$$

Let $k_{p1} + k_{i1} = k_0$, $\frac{k_{p1}}{k_{p1}+k_{i1}} = f_0$, then

$$C_1(z^{-1}) = k_0 \cdot \frac{z - f_0}{z - 1}$$

As mentioned above, one of the poles of the speed-to-fuel rack model is 0.9368. Let f_0 be a zero to cancel this slow pole $p_{1(1)}$, so

$$f_0 = p_{1(1)} = \frac{k_{p1}}{k_{p1} + k_{i1}}$$

The transfer function of the closed loop is,

$$H(z^{-1}) = \frac{k_0 b_{0(1)} z}{(z-1)(z - p_{2(1)}) + k_0 b_{0(1)} z}$$

The characteristic equation is,

$$(z-1)(z - p_{2(1)}) + k_0 b_{0(1)} z = 0$$

$$z^2 - (1 + p_{2(1)} - k_0 b_{0(1)})z + p_{2(1)} = 0$$

k_0 can be determined from desired characteristic equation, then k_{p1} and k_{i1} can be found by resolving the following equations,

$$\begin{cases} \frac{k_{p1}}{k_{p1} + k_{i1}} = f_0 \\ k_{p1} + k_{i1} = k_0 \end{cases}$$

The selection of the gains in controller 1, k_{p1} and k_{i1}, are based on three performance indices, rising time T_r, settling time T_{st}, and percent overshoot P. The rising time is the time required for the system to move from 10 to 90 percent of the steady-state value (before the first overshoot). The settling time is defined as the time when the speed error settles to be 5 percent of the step input. The percent overshoot is defined as

$$P = \frac{M_{pt} - 1}{1} / 100$$

$$M_{pt} = \frac{\text{peak speed value}}{\text{required speed}}$$

So the optimal criterion for the design of the speed loop is to minimize

$$J_1 = q_{1(1)} \cdot P + q_{2(1)} \cdot T_r + q_{3(1)} \cdot T_{st}$$

Fig. 2 shows the result of J_1 against the variation of k_0 which determines two gains in controller 1. Since the best performance of controller 1 in the speed loop in terms of speed control might not result in the best results for a combined speed and smoke control system, three groups of the gains of the controller 1 were selected for the design of the whole system. In Table 1 are the selected coefficients of controller 1 based on the model at engine speed 800RPM and torque 30 Nm.

Table 1 Selection of the coefficients in controller 1

No.	k_{p1}	k_{i1}	percent overshoot	rising time (seconds)	settling time (seconds)
1	3.75	0.25	0.0	0.60	1.50
2	4.68	0.32	0.0092	0.54	1.40
3	5.62	0.38	0.0221	0.40	1.60

Fig. 3 are the simulation results of the engine speed and smoke outputs with the controller in No. 3 of Table 1. The speed outputs with controller 1 in No. 1–3 of Table 1 are shown in Fig. 4. These results were obtained from the processes with only speed control.

Design of the combined speed and smoke control system

The design of the speed loop was to select the gains of the controller 1 with the emphasis put on canceling the slow pole in the transfer function of fuel rack-to-speed, and then meet the performance specifications. Instead of this method, the coefficients of the controller 2 for smoke error processing and the controllers of the combined speed and smoke control system were determined by minimizing the objective function J_2. Based on the optimization of engine speed performance, the requirement of the engine smoke output was adopted as one of the three performance indices. The other two indices were percent overshoot and settling time of the engine speed. The objective function was

$$\min J_2 = q_{1(2)} \cdot P + q_{2(2)} \cdot T_{st} + q_{3(2)} \cdot \sum e_{sm}$$

Controller 2 was chosen as a proportional controller. Still based on the models at engine condition of 800RPM and torque 30Nm, simulations are carried out with both controllers for processing speed and smoke signals. For the three selected PI controllers in Table 1, the gain of the controller 2 varied in a range to find the k_{p2} with which the objective function J_2 was minimum.

Fig. 5 was the results of J_2 against k_0 and k_{p2}. k_0 varies from 4.0 to 6.0 with 0.2 increased in each step, and k_{p2} varies from 0.1 to 2.0 with 0.2 increased in each step. As shown in the figure, the optimal value of k_{p2} is 1.2 when controller 1 has k_{p1} of 3.75 and k_{i1} 0.25.

EXPERIMENTAL RESULTS

The control systems designed above were tested to demonstrate that the speed and smoke control system was capable of reducing the smoke 'puff' at transients while the speed requirement was met.

Test conditions

Four speeds and three load settings were chosen to span the operating range. 700, 800, 900 and 1000RPM were chosen as the speed test points. For each test cycle, the engine ran at steady-state condition at first, one of the load points was set initially. After the required speed was input from the keyboard by the operator, the engine was accelerated by the control program from one speed point to another. The engine ran at a new speed-load condition until next engine speed was required. The speed output and the smoke output during the acceleration were sampled, converted and stored into the data files.

Results and discussion

Fig. 6–8 show the experimental results of the designed engine control system. In Table 2 are the corresponding engine test conditions of the figures. 'Speed' means the step of the required engine speed. 'Torque' means the initial load in each acceleration process. The limited smoke levels in the experiments were determined as the levels at steady-state engine condition.

Table 2 Engine test points

Fig.	Speed(RPM)	Torque(Nm)	k_{p1}	k_{i1}	k_{p2}
6	700-900	40	3.75	0.25	1.20
7	700-900	30	3.75	0.25	1.20
8	700-900	30	2.82	0.18	1.20

In Fig. 6, the smoke 'puff' (solid line) was produced by the control program which simulated the usual engine acceleration process without smoke detection. The doted line is the result when the requirement of engine speed was kept the same but with smoke feedback control.

In Fig. 4.16 and 4.17 are the results of controlled engine speed and smoke. They show that the smoke 'puff' is reduced with the control effects while the required speed steps are met at light load conditions.

The results shown above demonstrate that the smoke exhausts during engine transients could be reduced with the developed control system. The experimental results of this engine control system were satisfied when the engine acceleration spanned the range of 700–900RPM with light torque below 40Nm.

CONCLUSIONS

The feedback control of the engine smoke was realized using a new smoke sensor. The control system detected the smoke during engine transients. The controllers were designed with a method which combined the conventional PI controller design and the observation of the optimal engine performance. This design method was mainly used to solve the problems caused by the nonlinearity of the engine models. Experimental results at light-load engine conditions demonstrated that this engine control system effectively reduced the smoke 'puff' when it controlled the engine acceleration to meet the requirement of the engine speed.

REFERENCES

[1] Sell, J.A. and Chang, M.F., "Closed-loop control of an engine's Carbon Monoxide emissions using an Infrared Diode Laser", *SAE* 820388
[2] Cassidy, J.F., "A computerized on-line approach to calculating optimum engine calibration", *SAE* 770078
[3] Dohner, A.R., "Transient system optimisation of an experimental engine control system over the federal emissions driving schedule", *SAE* 780286
[4] Zanker, P.M. and Wellstead, P.E., "On self-tuning diesel engine governors", Control System Center Report No. 422, UMIST, 1978
[5] Mihele, W.P. and Citon, S.J., "An adaptive idle mode control system", *SAE* 840443
[6] Collings, N., Baker, N.J. and Wolber, W.G., "Real-time smoke sensor for diesel engine", *SAE* 860157
[7] Collings, N., Hong, G. and Baker, N.J., "Diesel smoke transient control using a real-time smoke sensor" *SAE* 871629
[8] Hong, G. and Collings, N. "Engine modelling for diesel smoke control", *SAE* 890389

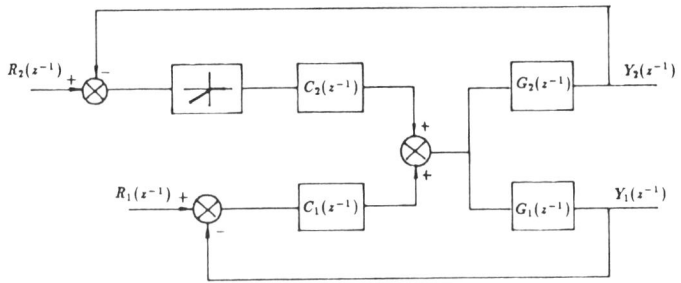

Fig. 1 Diagram of the engine smoke control strategy

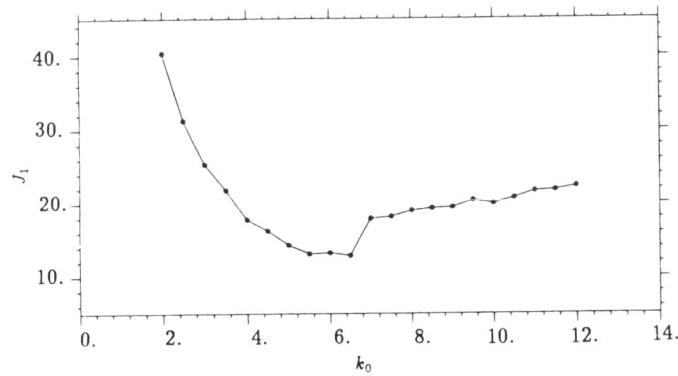

Fig. 2 Results of J_1 against the variation of k_0

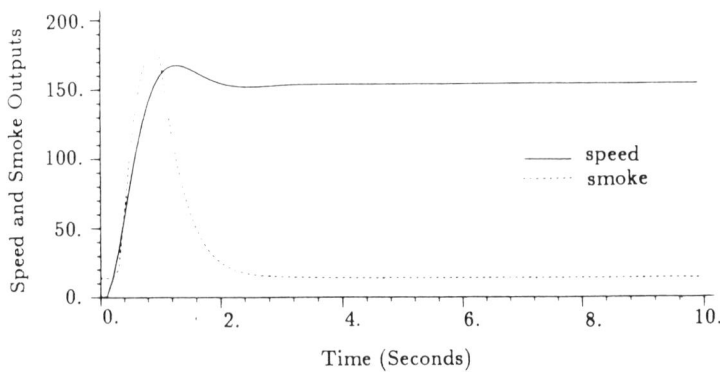

Fig. 3 Simulation results of the controller of No. 3 in Table 1

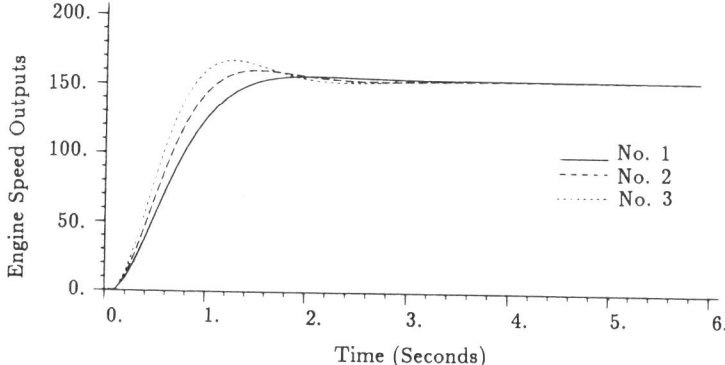

Fig. 4 Speed outputs of the controllers of No. 1-3 in Table 1

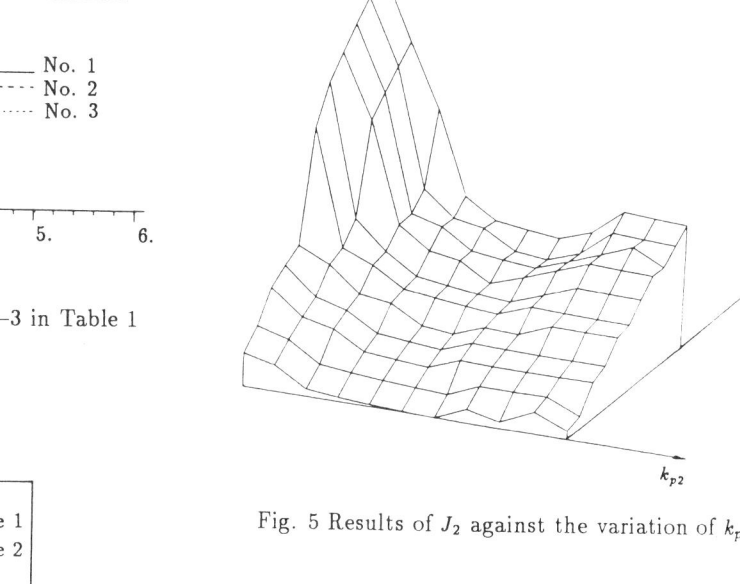

Fig. 5 Results of J_2 against the variation of k_{p2}

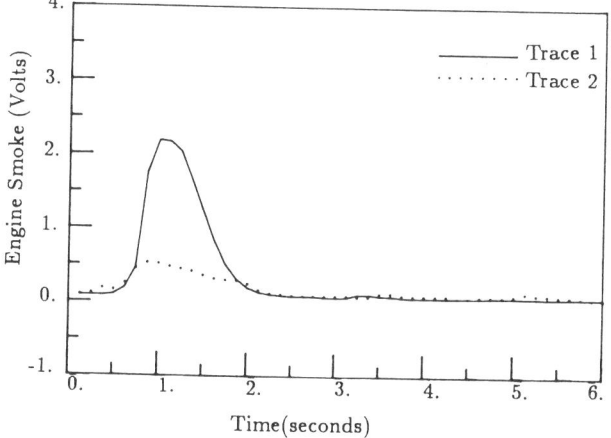

Fig. 6 Comparison of the engine smoke outputs with and without smoke control

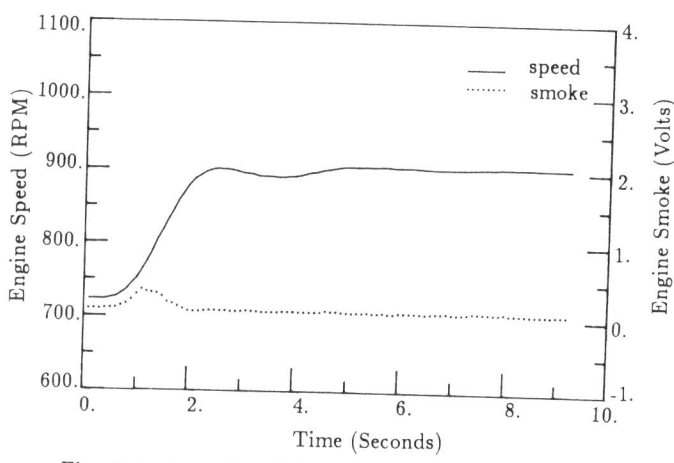

Fig. 7 Test results of the designed engine control system

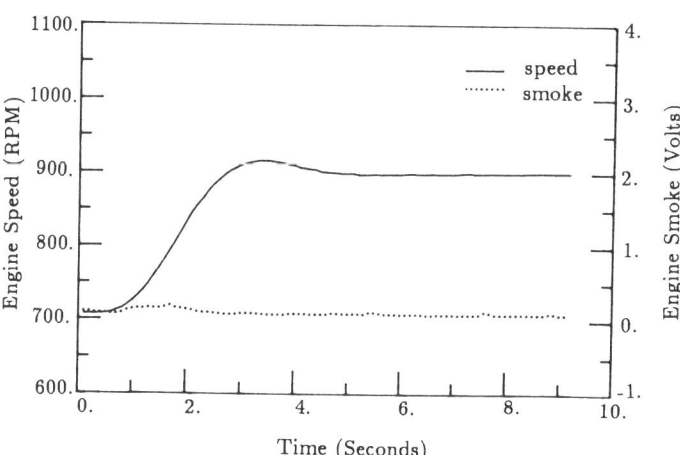

Fig. 8 Test results of the designed engine control system

890389

Engine Modelling for Diesel Smoke Feedback Control System Design

G. Hong and N. Collings
Engineering Dept.,
Cambridge Univ.,
England

ABSTRACT

A novel smoke sensor was used to measure the smoke response to the fuel rack on a diesel engine. The conventional modelling methods used for engine control were investigated. The synchronization technique and Recursive Least Square method were applied to engine modelling and two models for controller design were derived.

UP TO DATE CONTROL theory is being used more and more widely to satisfy emission requirements and fuel economy standards. Engine modelling is always the first step in the solution of a control problem. Generally, there are two ways to determine the mathematical model of an engine system:
1) mathematico-physical analysis based on general physical laws or process dynamics.
2) Experimental analysis based on the information obtained by measurements.

Reviews of the modelling technique used for engine control, with respect to the form of the input signals selected, show that frequency response technique has been mostly used. In 1974, Windett and Flower[1] applied sample-data theory to frequency response measurements on a large diesel engine. To eliminate the scatter produced due to the signal processing, a synchronizer was used to produce repeatable measurements and demonstrated by applying it to a variety of simple sampled-data systems. The input signal was a sine wave of frequency from 0.1 to 10Hz.

Chang and Sell[2] gave another example of applying this technique to engine modelling with an infrared diode laser spectrometer to find the dynamic effects of A/F ratio perturbations about stoichiometry on CO emissions and engine torque. In their work, the input signal was a biased sinusoidal voltage to perturb engine A/F ratio at different frequencies. The sinusoidal perturbation was varied from 0.1 to 10 Hz in ten steps per decade. Two models were shown, one was for A/F effect on torque, another was for correlation of engine torque and CO emissions.

Flower et al did more work on engine modelling using P.R.B.S. techniques [3] to get the transfer function of engine speed to fuel rack position. The P.R.B.S. pulse rate and sample time were synchronized with the engine firing. But details of the model were not discussed.

System dynamic modelling with an identification algorithm was shown in[4], where Morris et al used physical principles to obtain a block diagram of the system, and then Landau's hyperstable identification algorithm to determine the parameters of the model. Based upon a sampling interval of one engine revolution, the response of the torque to throttle was described by a fourth-order dynamic model.

The ability to model an engine suggests that modern control such as system analysis and design technique can be employed to the engine control problem. However, this has been impeded by the absence of suitable dynamic models. And the basic reason is the lack of the sensors. This paper shows how a novel smoke sensor was used as a basic component to get the necessary information for modelling. The problem of using conventional modelling techniques will be discussed at first, and then the application of the improved identification method to two models: speed-to-fuel rack and smoke-to-fuel rack will be described.

APPLICATION OF FREQUENCY TECHNIQUE FOR ENGINE MODELLING

In frequency domain, the transfer function $G(s)$ of a linear system can be shown to be a ratio of two polynomials in s, each of which can be factorized to give

$$G(s) = \frac{Y(s)}{U(s)} = \frac{k(s-z_1)(s-z_2)....(s-z_m)}{(s-p_1)(s-p_2)....(s-p_n)}$$

If a sine wave of unit amplitude and of frequency ω rad/second is taken as the input to system:

$$U(t) = \sin\omega t$$
$$U(s) = \frac{\omega}{s^2 + \omega^2}$$

The output y(t) will be

$$[y(t)]_{t \to \infty} = |G(j\omega)|sin(\omega t + \angle G(j\omega)) \quad (1)$$

The magnitude and phase as functions of frequency can thus be obtained from the transfer function $G(s)$ by replacing s by $j\omega$ and determining the modulus and argument of $G(j\omega)$, which for any particular frequency is generally a complex number.

To model an engine system, the output of engine speed and movement of the fuel rack are continuous, however, the actual injector contains a pure sampling mechanism.

Engine speed or smoke depends on the behavior of the engine fuel supply. For a typical fuel injector system of a diesel engine, the injection time is of the order of 5 percent of engine cycle time, so the injection can be treated as sampling the movement of the fuel rack. The frequency of the input sine wave must be at least twice the frequency of the engine firing, otherwise, 'aliasing' will happen and the response of the output to the input will never correspond to equation (1).

To be able to determine the dynamic characteristics of the operating process of an engine, it is necessary to determine the value of the maximum acceptable frequency of the input sine wave from the engine firing rate to avoid aliasing problems. At low engine speed, the sine wave is limited in very low frequency, for example, about 3 Hz at an engine speed of 700RPM. In practice, the Bode diagram starts to become scattered when the frequency of the input sine wave is above 1/4 of the engine firing rate.

MODELLING METHOD TO BE EMPLOYED

Based on the review of the modelling techniques used on engine control, the analysis of the problem to apply frequency response method and consideration of the digital form of the transfer function to be required in this control system, parameter estimate was chosen as the modelling method.

This method consisted of
1) Using P.R.B.S. as the input signal which acted on the fuel rack;
2) Utilise synchronization technique to acquire the data for identification;
3) Estimating the parameters of the models with Recursive Least Square method.

EXPERIMENTAL

The control objective was to reduce or remove the 'smoke puff' during the engine transients. Models for two pairs of input-output were expected for this control system, speed-to-fuel rack and smoke-to-fuel rack.

The effectiveness of such a control system was discussed in [5] which shows how the smoke 'puff' can be removed. Here we are attempt to model the engine more exactly and improve the control performance.

Tests for modelling were performed on a AEC single cylinder, DI diesel engine. The component of the experimental set-up are described as following:
1) Actuator: The fuel rack position is altered by a servomotor which is driven by a command signal from the computer.
2) Sensors: A speed signal is taken from a tachogenerator on the crankshaft. The smoke sensor is described in [5] and [6]. The sensor is inserted into the exhaust stream and gives out a signal due to the charged nature of some of the particles in the exhaust[7]. Calibration tests comparing this system with a Hartridge opacity meter[6] gave a correlation coefficient of 0.87 which confirms a good agreement between the sensor's output and Hartridge Units.
3) The computer PDP/11-23 PLUS was used for data acquisition for modelling and complementation of the controller in later work.
4) The controlled engine: AEC single cylinder, DI diesel engine.

So, the input of the system model is the signal to drive the actuator which moves the fuel rack, and the outputs are engine speed and smoke.

MEASUREMENT OF THE INPUT AND OUTPUTS

The diagram of the system for modelling is shown in Fig.1 and the layout for data acquisition in Fig. 2.

1) Generation of P.R.B.S. signals

The P.R.B.S. signal was generated by seven shifters to get a 127 bit sequence. The amplitude of the signal was selected by the user before the P.R.B.S. was generated.

A program linked with an assembly language routine was used to sample and store the data results of the input and outputs into data files, and send out the P.R.B.S. analogue signal to the servo motor. The program read in the P.R.B.S. data sequence, added it to a basic value(the basic voltage to keep the steady-state engine speed), then the assembly language routine moved each of the data sequences from the memory to the D/A converter when the engine trigger was enabled. This analogue signal then moved the servo motor and this signal level was held until next trigger's arrival. 300 samples were taken for each program execution.

2) Synchronization of sampling and engine rotation

As mentioned above, the diesel engine is a typical sampler. When the signals of the system are sampled, it is required that the sampling of the continuous parts of the system occurs at regular intervals of the same variable that governs the natural samplers. To meet this requirement, the trigger starting the D/A converter and sending out the P.R.B.S. pulse had the same frequency as engine firing. The time interval to keep one P.R.B.S. pulse was chosen to be the time for one engine cycle. Each trigger to P.R.B.S. pulse happened in the exhaust process, this made the position of the fuel rack unchange during injection period.

The diagram of this synchronization is displayed in Fig. 3. As shown there, the sample frequency of the input and outputs is the same as engine revolution. This means that sample frequency is twice of that of P.R.B.S. input. The A/D conversion happens at the time when the piston arrives at TDC.

3) Selection of the amplitude of the P.R.B.S signals.

The amplitude of P.R.B.S. was selected to make a small perturbation on the outputs when the data for identification were acquired. The final model derived was in digital form whose discrete time interval was constant. In fact, however, the real sample frequency for data acquisition was varied because the engine speed was changing due to the perturbation on the fuel rack and the sampling was synchronized with the engine revolution. Thus the fixed sample frequency in the identified model was actually an average of the real sample frequencies. The more the engine speed was perturbated, the more the sample frequency varied. The accuracy of the model would be unacceptable if the sample frequency in the test changed in a wide range.

The engine speed was sensitive to the small variation of the fuel rack's movement at low frequency, but the smoke signal could not be excited when the P.R.B.S. pulse is smaller than 0.25V. Hence, the amplitude of P.R.B.S. input was chosen to be higher than 0.25V. basically and with different gains to investigate the non-linearity of the system.

SYSTEM IDENTIFICATION ALGORITHM

Recursive Least Square(RLS) method was used to estimate the parameters of the models.

MODEL IN RLS —- The model underlying RLS is

$$A(q^{-1})y(t+1) = B(q^{-1})u(t-nd) + C(q^{-1})v(t+1)$$

where

y — the output

u — the input

v — uncorrelated, zero-mean random sequence or white noise

$A(q^{-1}) = 1 + a_1 q^{-1} + a_2 q^{-2} + \ldots + a_{na} q^{-na}$

$B(q^{-1}) = b_0 + b_1 q^{-1} + b_2 q^{-2} + \ldots + b_{nb-1} q^{-nb+1}$

$C(q^{-1}) = 1 + c_1 q^{-1} + c_2 q^{-2} + \ldots + c_{nc} q^{-nc}$

q^{-1} — represents the unit delay operator,

$q^{-1} y(t+1) = y(t)$

ESTIMATE EXECUTE — The parameter estimate $\hat{\theta}(t)$ based on t data point is given by

$$\hat{\theta}(t) = \hat{\theta}(t-1) + W(t-1)[y(t) - \hat{y}(t)]$$

where

$$\hat{y}(t) = \phi(t-1)\hat{\theta}(t-1)$$

$$W(t) = \frac{\mathbf{P}(t-1)\psi(t-1)}{\psi(t)\mathbf{P}(t-1)\psi'(t) + \lambda(t)}$$

$$\psi(t) = \frac{1}{\hat{c}(q^{-1})k(t)}\phi(t)$$

where k(t) is a contraction factor and $\lambda(t)$ is a forgetting factor. a_λ and a_k control the rates at which $\lambda(t)$ and k(t) reach their final values. $0 \leq a_\lambda \leq 1, 0 \leq a_k \leq 1$

$\mathbf{P}(t)$ is a scaled version of the covariance of the estimate given by

$$\mathbf{P}(t) = \left[\mathbf{P}(t-1) - \frac{\mathbf{P}(t-1)\psi'(t)\psi(t)\mathbf{P}(t-1)}{\lambda(t) + \psi(t)\mathbf{P}(t-1)\psi'(t)}\right]/\lambda(t)$$

PARAMETER ESTIMATE

Parameter estimation was completed using Matrixx, a proprietary software package for control system design and analysis.

Parameters of the models were estimated with 50-100 sample points in each engine test condition. The initializations of the parameters was zero. The convergency of the parameter estimate is shown in Fig. 4 and Fig. 5. As shown in the figure, there are some fluctuations in the first 4-20 iterations of the parameter estimation for speed models. Then the values of the parameters settle down to steady-state. There are more fluctuations in the estimation for the smoke model and the values of the parameters do not become as stable as those of the speed model.

Time delay of the models were determined from the sampled data. The time delay of speed-to-fuel rack was one sample interval and smoke-to-fuel rack two intervals.

The order of the speed-to-fuel model was assumed to be first order by referring to the mathematical model of engine and the experimental results in previous work[8]. For an engine on which loading is unchanged, the transfer function between fuel rack and speed is nearly a pure integrator if the engine is not governed, and the fueling is not rich.

Little previous work appears to have been published on the investigation to smoke-to-fuel model. In the work of Chang and Jeffery[2], the model of fuel rack-to-CO emissions was derived as a second order one. Since the relative proportions of particulates, unburnt HC and CO might be conceived to be roughly in proportional to each other, as they are all to some extent caused by oxygen deficiency, the transfer function of smoke-to-fuel rack was assumed to be second order. This assumption may be seriously in error however, and a better data is required.

Also, a range of the model's orders were investigated and statistical measures, accuracy of the models were utilized to finally determine the order of the model. The coefficients produced for the various models are given in Table 1 and 2 respectly for speed and smoke models. In Table 1, the data were taken at engine steady-state speed 800RPM and torque 30 Nm, and Table 2, speed 800RPM and torque 60 Nm. The speed model which consistently gives the minimum square error is that for $na=2$, $nb=0$ and $nd=1$. The smoke model agrees with the assumed one which has $na=2$, $nb=0$ and $nd=2$. Two tables show that the error tends to be bigger with increasing order.

Similar results about the order of the transfer function were obtained at other speed-torque points, here only the models derived are given in Table 3 and 4.

Table 1 Coefficients for various orders of speed-to-fuel rack model

na	nb	nd	a_1	a_2	a_3	b_0	b_1	b_2	V
1	0	1	-0.9318			0.0296			1830.2
2	0	1	-1.5375	0.5627		0.0145			1104.3
2	1	1	-1.4198	0.4621		0.0136	0.0067		1071.9
3	0	1	-1.6712	0.8033	-0.1009	0.0131			1472.6
3	1	1	-1.4159	0.4888	-0.0305	0.0131	0.0076		1122.2
3	2	1	-1.4328	0.4983	-0.0249	0.0127	0.0080	-0.0012	1168.7

Table 2 Coefficients for various orders of smoke-to-fuel rack model

na	nb	nd	a_1	a_2	a_3	b_0	b_1	b_2	V
1	0	2	-0.8798			0.0476			372.297
2	0	2	-0.5719	-0.2883		0.0616			226.351
2	1	2	-0.4595	-0.3861		0.0653	0.0036		223.151
3	0	2	-0.6330	-0.1793	-0.0524	0.0603			224.915
3	1	2	-0.3700	-0.1668	-0.2788	0.0674	0.0203		215.791
3	2	2	-0.2769	0.1794	-0.6517	0.0708	0.0318	0.0173	223.244

Table 3 Parameters of speed-to-fuel rack model at different engine conditions

Speed (RPM)	Torque (Nm)	Gain k_1	Pole 1 p_{11}	Pole 2 p_{12}	Time delay $nd1$ (sample interval)
800	30	0.0145	0.9368	0.6006	1
800	60	0.0094	0.9447	0.5369	1
700	30	0.0140	0.9613	0.5741	1
700	72	0.0075	0.9775	0.4801	1
900	30	0.0180	0.9351	0.3583	1
900	60	0.0101	0.9556	0.3807	1

Table 4 Parameters of smoke-to-fuel rack model at different engine conditions

Speed (RPM)	Torque (Nm)	Gain k_2	Pole 1 p_{21}	Pole 2 p_{22}	Time delay $nd2$ (sample interval)
800	30	0.0099	0.7767	0.1197	2
800	60	0.0335	0.9506	0.3536	2
900	45	0.0616	0.8943	-0.3224	2
900	60	0.0322	0.9448	0.2132	2

DISCUSSION

The engine speed response to the fuel rack movement was tested at different speed-torque points. The results in Table 3 shows that in the speed model there is always a slow pole which is bigger than 0.9 and less than 1 in Z domain, and the gain between output and input decreases with the increasing of the engine torque. A tendency can be found that the one of the poles in the transfer function goes towards the position more unstable when the speed is lower, this agrees with the results of engine modelling in the previous work.

The poles in the smoke model are more stable than that of the speed model. The gain of the transfer function increases very quickly with the increasing of engine torque. However, it has been found that smoke model was very difficult to identify, due mainly to its nonlinearity. Different sizes of the P.R.B.S. input were used to observe the nonlinearity of the smoke model at the same speed-torque test point. The steady state gain changed considerably with the variation of the size of the P.R.B.S. signals. From the control standpoint, the main objective to be controlled is engine speed and the smoke is to be detected in the transients, so the requirements to the speed model and smoke model are different and they can be accepted with different accuracy.

The time delay of either speed response to fuel rack or smoke is less than on engine cycle from the view of engine operation. In the models shown in Table 3 and 4, the time delay of the speed model is one sample time interval and smoke model two sample time intervals. In fact, the time delay results from the lag in the engine process and the sampling action of the control system. During the time one P.R.B.S. pulse takes, the first sample starts at the TDC after compression stroke, the second one happens at the TDC after expansion stroke, at this moment the speed signal affected by the alteration of the fuel supply is ready to be output but the smoke signal is being processed. So the smoke signal can only be sensed one sample interval later than the speed signal.

Fig. 6 and 7 are the simulation results showing the accuracy of the models derived. The input for simulation was taken still the P.R.B.S. used for identification, and the simulated outputs were compared with the actual outputs. The agreement between the measured and calculated results is quite acceptable and comparison shows that the transfer function of speed-to-fuel rack is more accurate than that of smoke-to-fuel rack.

CONCLUSIONS

A new smoke sensor makes it be realizable to measure the diesel smoke response to the fuel rack with an experimental method.

The work in this paper was undertaken to utilize the modelling techniques, and taking proper account of the sampling nature of the diesel engine. The Recursive Least Square method was employed to identify the transfer functions of fuel rack to engine speed and fuel rack to engine smoke.

The engine speed model can be linearized as a second order transfer function in Z domain with a time delay of one sample interval, i.e. one engine revolution. The nonlinearity of the smoke model was investigated in a preliminary way. Identified in a certain range, the smoke model was approximated as a second order transfer function with a time delay of two sample intervals.

Simulation work shows that the accuracy of the smoke model is lower than that of the speed one. More work could be suggested on the non-linear model of smoke-to-fuel rack for more accurate engine smoke control in the future work.

REFERENCES

[1] Windett, G.P. and Plower, J.O., "Sampled-data frequency response measurement of a large diesel engine", *Int. J. Control*, Vol. 19, No. 6, 1974, pp. 1069-1086

[2] Chang, M. and Sell, J.A., "A linear model of engine torque and Carbon Monoxide emissions", *SAE* 830427

[3] Flower, J.O. and Hazell, P.A., "Sampled-data theory applied to the modelling and control analysis of compression-ignition engines – Part II", *Int. J. Control*, 1971, Vol. 13, No. 4, pp. 609-623

[4] Morris, R.L., Borcherts, R.H., Warlick, M.V. and Hopkinst, H.G., "Spark ignition engine model building – an identification approach to throttle-torque response", *Int. J. of Vehicle Design*, Vol. 3, No. 1, 1982

[5] Collings, N., Baker, N. and Wolber, W.G., "Real-time smoke sensor for diesel engine", *SAE* 860157

[6] Collings, N., Hong, G. and Baker, N.J., "Diesel smoke transient control using a real-time smoke sensor", *SAE* 871629

[7] Kittelson, D.B. and Collings, N., "Origin of the response of electrostatic particle probes", *SAE* 870476

[8] Haddad, S. and Watson, N. (Ed.), *Principles and Performance in Diesel Engineering*, Ellis Horwood Ltd, 1984

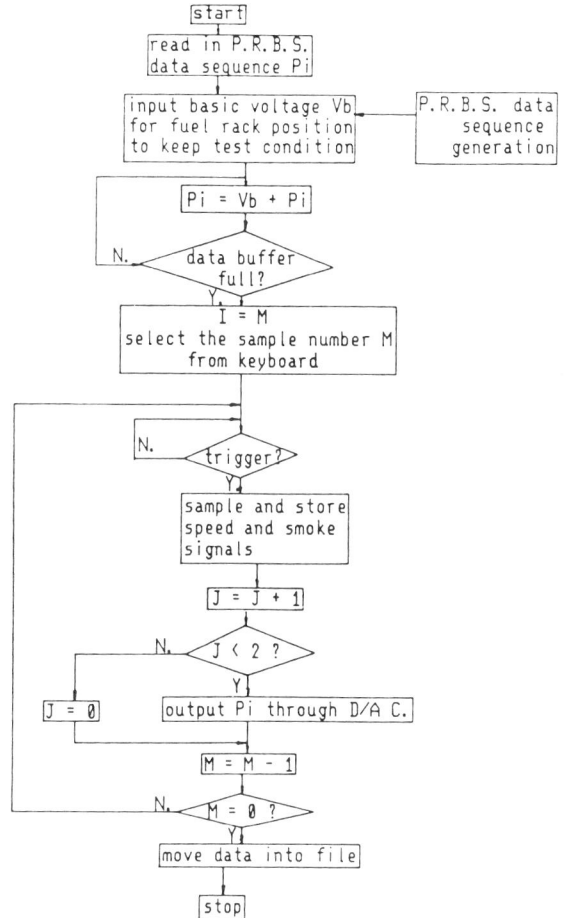

Fig. 2 Flow chart of data acquisition

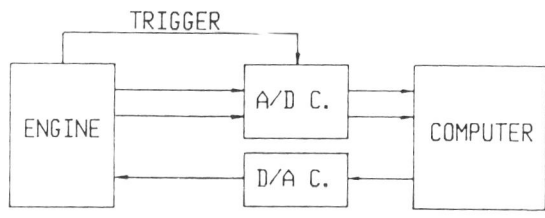

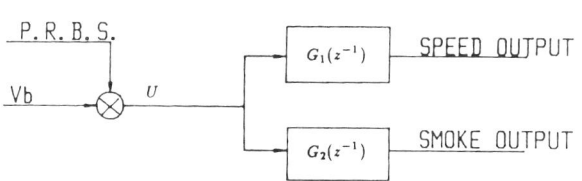

Fig. 1 Diagram of the modelling system

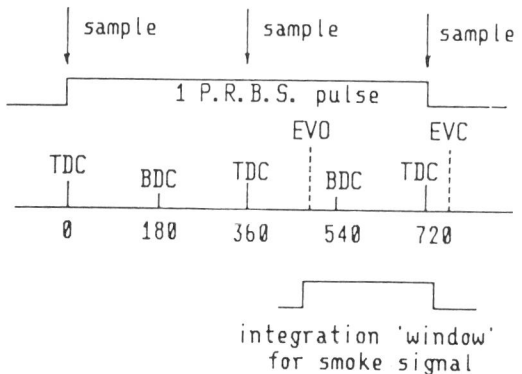

Fig. 3 The event which occurs with synchronization technique

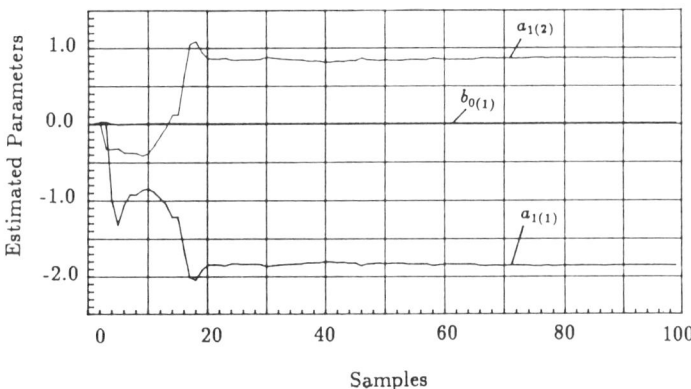

Fig. 4 Convergence of the parameter estimates (speed model, second order)

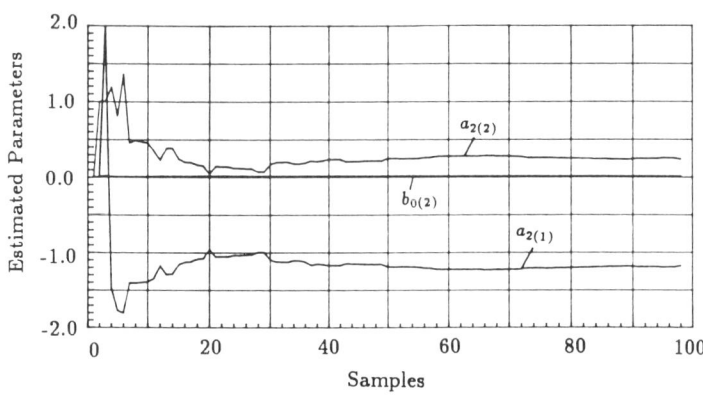

Fig. 5 Convergence of the parameter estimates (smoke model, second order)

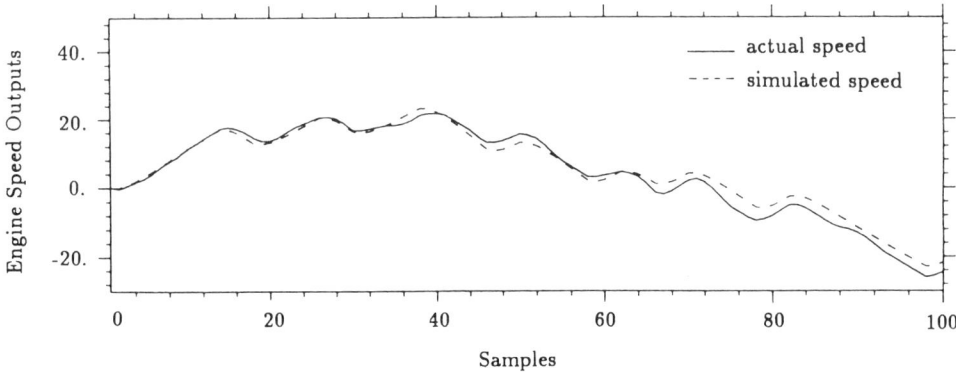

Fig. 6 Comparison of real simulated speed outputs (speed=900RPM, torque=30Nm)

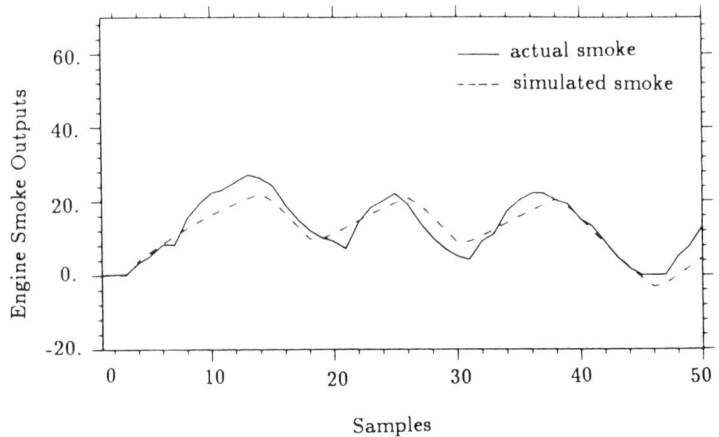

Fig. 7 Comparison of real simulated smoke outputs (speed=800RPM, torque=60Nm)

890391

Electronic Controls for John Deere Diesel Engines

Marvin K. Farr
Engine Engineering
John Deere Engine Group

ABSTRACT

Electronic engine control systems have been developed for the John Deere 400 Series 7.6L and 500 Series 10.1L diesel engines. These full-featured, microprocessor based systems provide significant advantages over mechanically controlled engines in on-highway, off-highway and stationary engine applications. Unique controller features include a timed power boost mode, transient smoke control without manifold sensors, switch selectable torque curves and speed regulation, special throttle options, adaptive throttle calibration, temperature compensation and factory "end-of-line" programming.

THE DEVELOPMENT OF ELECTRONIC CONTROLS for John Deere diesel engines was begun in the late 1970's. The initial emphasis of the program was to provide a control system which could provide significant benefits to the major users of Deere engines - traditionally off-highway. Thus the objectives included providing a highly reliable system for use in the rugged off-highway environments while at the same time being very cost competitive.

Some features were developed to meet specific Deere vehicle requirements and provide new enhancements. Other potential functions were eliminated to reduce the cost. A key criteria used in selecting the final complement of features was the desire to take advantage of the inherent flexibility of microprocessor based controls in adapting to new applications.

SYSTEM DESCRIPTION & COMPONENTS

The Deere engine control system is adapted as an option to the 400 and 500 Series diesel engines. The system includes the injection pump with a special electronic actuator assembly, an electronic engine control unit, several special sensors and the interconnecting wiring harness.

SYSTEM DESCRIPTION - Figure 1 is a diagram of a typical overall system installation. The electronically controlled inline injection pump utilizes the same basic hydraulic pumping mechanism used in mechanically governed inline pumps. The mechanical governor mechanism is replaced with an electronic actuator assembly. The usual throttle lever mechanism is removed and its function is implemented by a throttle position sensor input to the chassis-mounted engine control unit(ECU). The ECU controls both the fuel limiting for torque curves and the governing for speed control. A fuel temperature sensor and a fuel shutoff solenoid are installed at the fuel inlet to the injection pump. A special wiring harness interconnects the ECU, the injection pump and the various external control system inputs and outputs. The wiring harness also includes special ports for service.

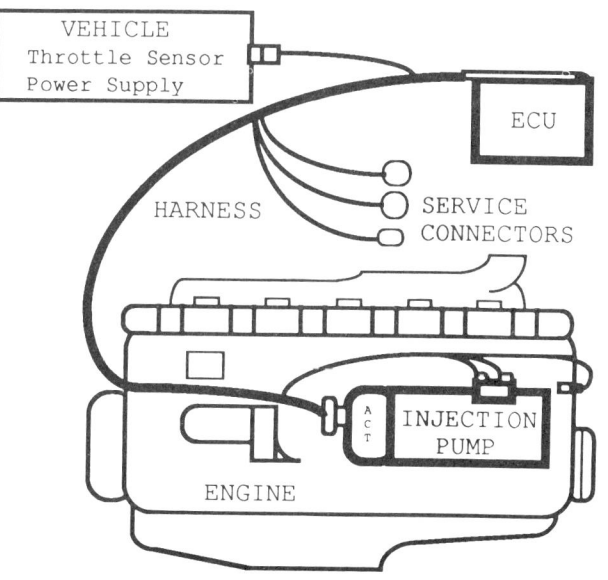

FIG. 1 - Control System Diagram

THE ACTUATOR ASSEMBLY - Figure 2 shows a sketch of the actuator assembly. The control rack [A] in the injection pump is spring-loaded to the zero stop (fuel shutoff) position. The linear solenoid [B] positions the fuel control rack in opposition to the return spring [C] force. Therefore, whenever power is disconnected (key OFF), the fuel is automatically shut off. The drive current to the linear solenoid is controlled by the engine control unit (ECU). The rack position sensor [D] provides positive feedback of the actual rack position to the ECU. This feedback sensor helps to insure a high degree of accuracy in the positioning of the control rack and also provides a means of detecting possible problems or failures in the control rack positioning system. The speed sensor [E] and toothed speed wheel [F] provide speed feedback to the ECU for controlling engine speed.

FUEL SHUTOFF SOLENOID - A special normally-closed fuel shutoff solenoid located at the fuel inlet to the injection pump provides a back-up shutdown system. The ECU can automatically interrupt power to the shutoff solenoid and force an engine shutdown if a critical fault condition is detected and eliminate the possibility of engine overspeeding. The shutoff solenoid is also controlled by a normal or emergency key-OFF operation.

FUEL TEMPERATURE SENSOR - A temperature sensor is located in the inlet fuel stream to monitor the fuel temperature. The ECU can be programmed to compensate the fuel delivery as the fuel temperature varies to maintain constant power output. The fuel temperature is also used to select an optimum starting sequence.

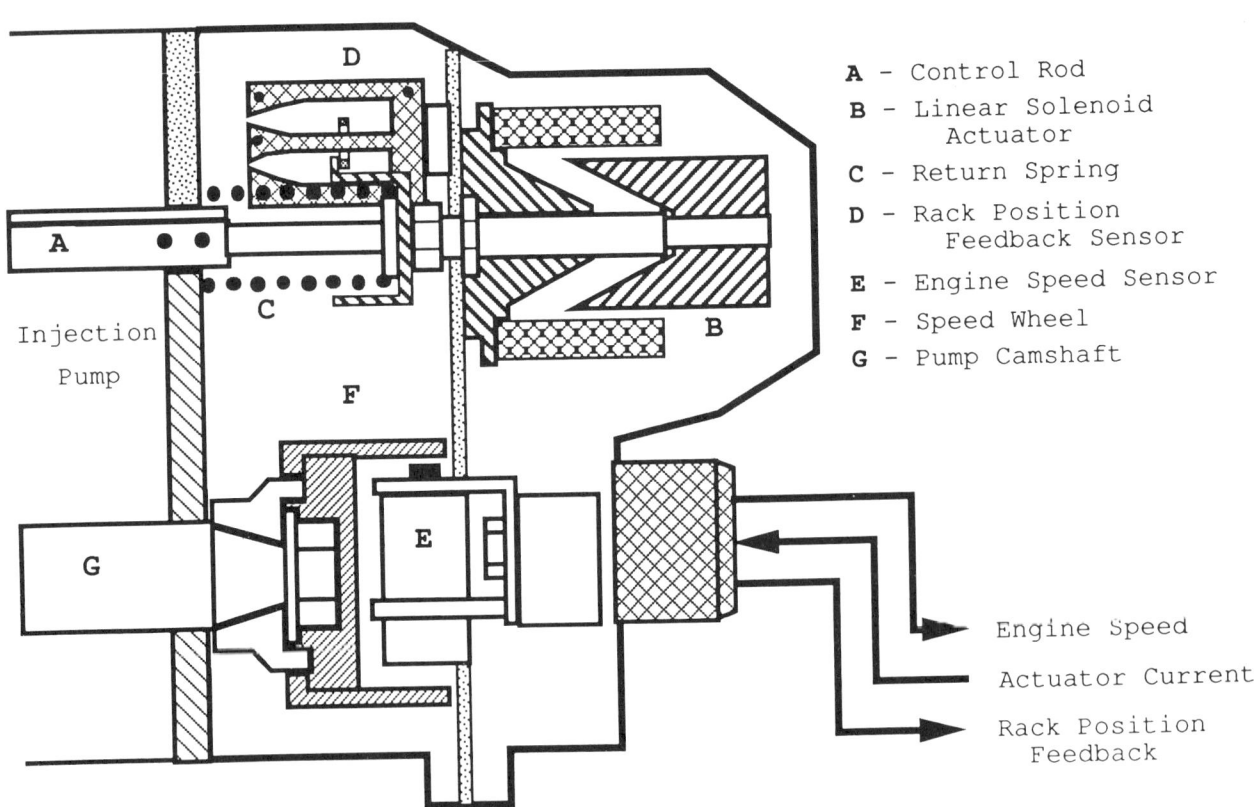

FIG. 2 - Electronic Acutator

ENGINE CONTROL UNIT (ECU) - The heart of this system is the engine control unit (ECU). The diagram in Figure 3 shows the various input and output (I/O) ports of the ECU. The functions associated with these I/O ports are determined by the ECU software and are discussed in the Control System Functions section of this paper. The ECU's were developed jointly with the injection system suppliers according to Deere specifications. The specific ECU and actuator hardware was designed and is manufactured by the pump suppliers. This helps to insure the complete compatibility of the controls with the rest of the injection system while meeting Deere requirements for functions and reliability. These microprocessor based units utilize hybrid circuits, EEROM's, EPROM's, etc. to provide cost effective, reliable and flexible controllers. The ECU's are designed to be chassis mounted and connected to the injection pump via the special wiring harness.

THROTTLE POSITION SENSOR - The throttle position signal is the primary operator input to the control system. The ECU provides a reference voltage for use with potentiometer throttle sensors. The potentiometers can be controlled by various types of hand throttles or a special foot pedal assembly. Several other throttle options will be discussed in following sections.

WIRING HARNESS - The wiring harness which interconnects the engine control system is a critical component of the system. Careful attention was given throughout the system development to verify that connectors were reliable and durable for all anticipated applications. In most installations, the user will provide the wiring harness. Therefore detailed wiring specifications and guidelines were developed to help insure system integrity.

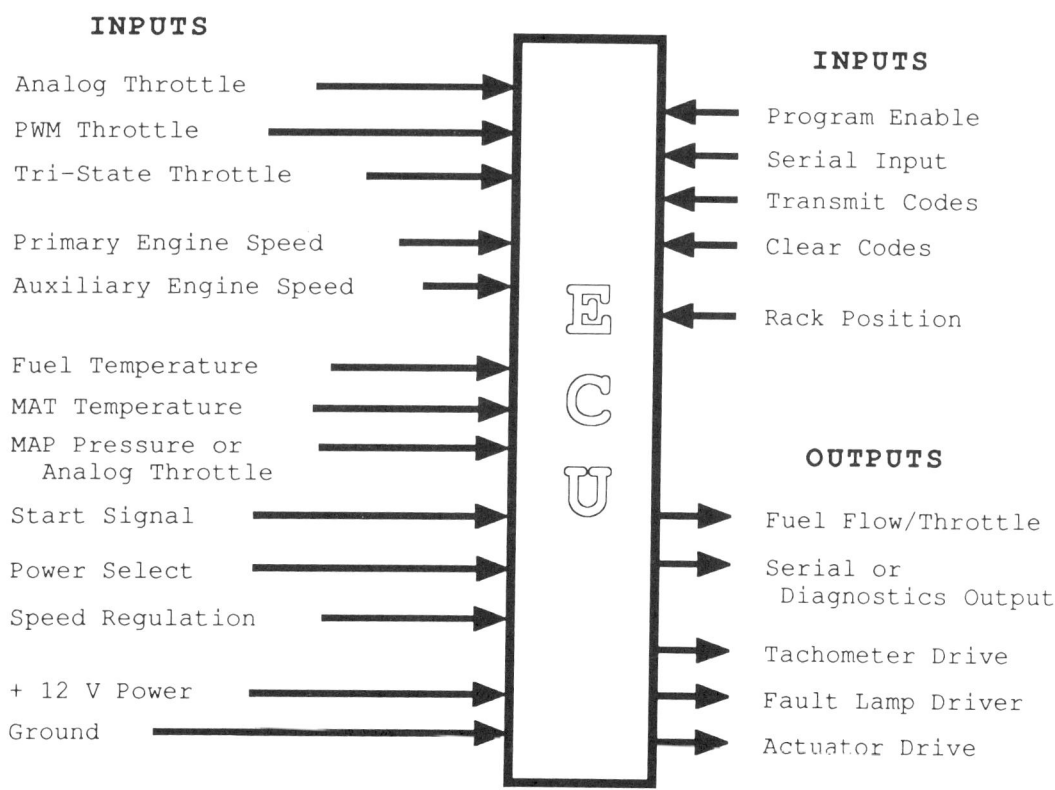

FIG. 3 - Controller Input/Output Ports

BASIC CONTROL SYSTEM FUNCTIONS

The control systems provide accurate control of the fuel quantity during all phases of the engine operation. This includes maximum fuel control (or torque curves), speed regulation or governing, air-to-fuel ratio (or smoke) control and special starting controls.

MAXIMUM FUEL QUANTITY CONTROL - The ECU is programmed to limit the maximum fuel delivery as a function of the engine speed. These limiting values can be programmed to obtain nearly any desired torque-speed characteristics within the engine/fuel system capabilities. These characteristics are highly repeatable since they are digitally programmed and are not dependent on any mechanical adjustments.

A tri-state power select input allows switching between any one of three preprogrammed torque-speed curves. One of the three curves has a timer associated with it which can limit the time of operation in this mode. This permits the use of a time-limited power boost option. Figure 4 shows an example of three torque curves wherein the curve labeled "N" is used during normal operations, curve "B" for a timed power boost mode and curve "D" for a derated mode.

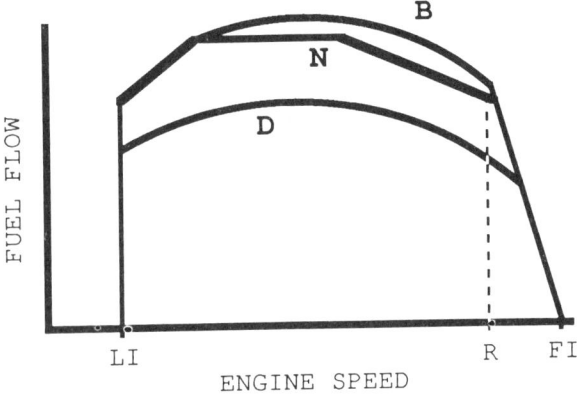

FIG. 4 - Maximum Fuel (Torque) Curves

GOVERNING CONTROL - The controller can provide either true all-speed governing or min-max governing with isochronous low idle speed control. The degree of speed regulation (droop) can be programmed to meet any application requirements including isochronous governing. Governor parameters can be readily optimized for each specific application.

The **all-speed governing mode** is shown in Figure 5 wherein the ECU controls the engine speed based on the percentage of full-throttle command from the throttle position input and the programmed speed regulation. The low idle speed can be isochronously governed independent of the speed regulation programmed over the rest of the operating range. This isochronous low idle control provides a fixed low idle speed independent of parasitic installation loading and improves low speed load-starting capability.

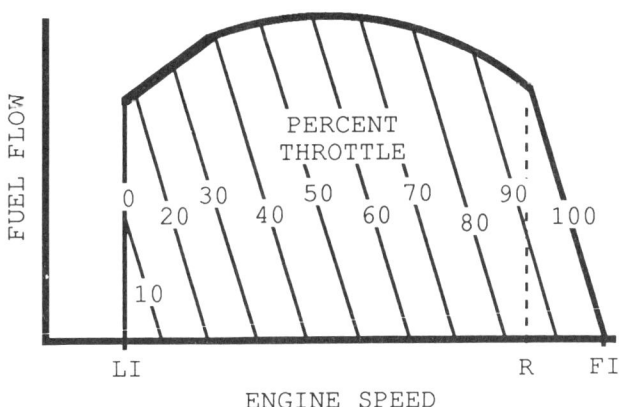

FIG. 5 - All-Speed Governing

In the **min-max governing mode**, the ECU controls fuel delivery via a throttle position versus engine speed map as shown in Figure 6. This allows the driveability to be tailored to the application. The minimum and maximum speeds are always regulated with the programmed droops similar to the all-speed governor mode.

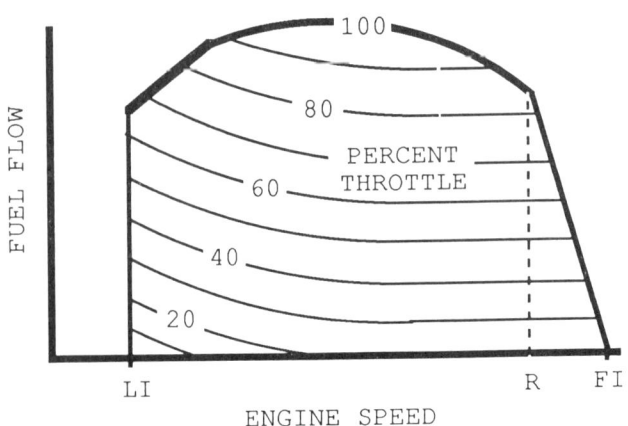

FIG. 6 - Min-Max Governing

A tri-state, speed regulation select input allows switching between any one of three programmed rated speed/speed regulation combinations. The speed regulation can be changed "on-the-go" to match a particular operation if desired. Road speed limiting and similar options are also implemented with this switch input.

Since the low idle, fast idle and rated speeds are determined by the digitally programmed ECU, these values are precisely repeatable. Thus unit-to-unit variations and idle speed adjustments are eliminated.

SMOKE CONTROL - The engine controller provides two techniques for smoke control. The first technique utilizes the common method of measuring the intake manifold air pressure (MAP) and optionally the manifold air temperature (MAT). The ECU then calculates the air density which is used to determine the fuel rate limit needed to control the smoke.

The engine controllers also provide an algorithm to control transient smoke based on previously determined engine characteristics. This technique actually provides better smoke control during transients than the sensor dependent methods since it is not subject to the errors and lag involved in air density measurements. A further advantage of this method is the elimination of sensors. This technique facilitates the optimization of the smoke control versus response characteristics for each application. Figure 7 shows a typical example of the smoke control during an engine acceleration using this mathematical model.

STARTING CONTROL - The controller utilizes the fuel temperature input to select an optimal starting sequence. Figure 8a shows an example of how the control rack positions are selected for various temperature ranges. This permits using excess fuel and retarded injection timing for cold starting while reducing the fueling for hot starts to eliminate black smoke. The ECU varies the control rack positions during starting as a function of the engine speed as shown in Figure 8b. This control sequence reduces the white smoke and faltering on cold starts, minimizes overshooting of the desired operating speed, reduces black smoke during hot starts and provides faster overall starting. The throttle position input is ignored by the controller until the starting sequence is exited. The ECU also provides a temperature dependent, increased low idle speed for a preset time after a cold start to reduce warm-up time.

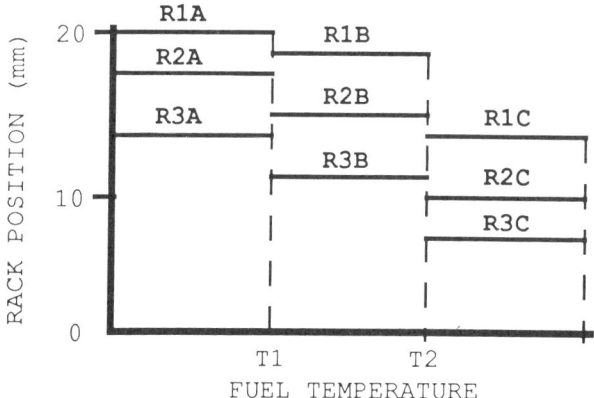

FIG. 8a - Starting Rack Positions

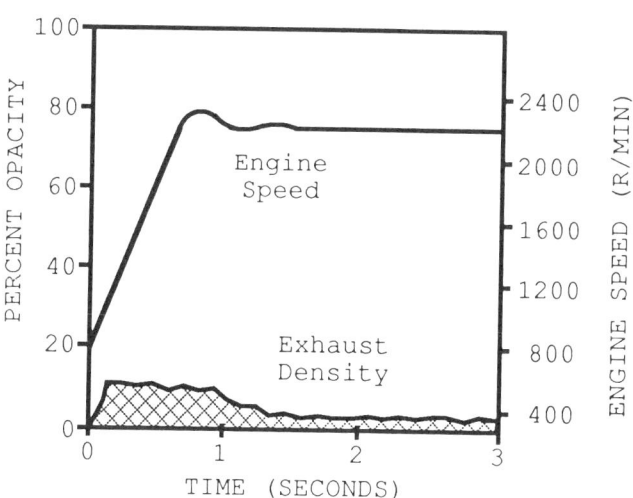

FIG. 7 - Acceleration Smoke Control
(Slow Idle to Fast Idle)

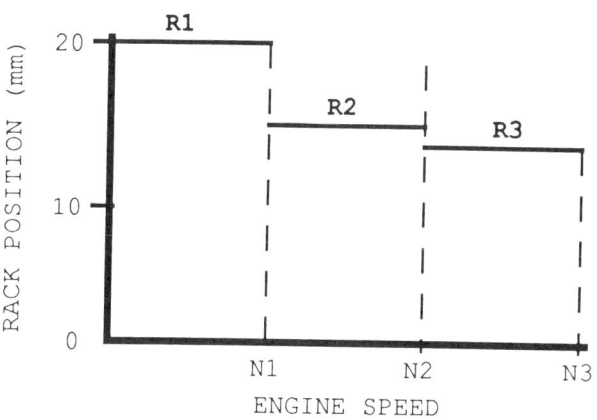

FIG. 8b - Starting Sequence

SPECIAL FEATURES

In addition to the basic control functions there are a number of other special features available for use in various applications. These include several throttle options, compensation for fuel temperature variations, a special fuel flow and throttle output signal, sophisticated self-diagnostics, a tach drive output and factory "end-of-line" programming.

THROTTLE OPTIONS - There are three basic throttle options plus combinations of these options available for use with the engine control systems. The most commonly used option is a standard analog throttle, potentiometer input. With the analog throttle, the throttle command is directly proportional to the voltage input level. The ECU can be programmed to accept two analog throttle inputs and will automatically scale the second input to the remaining travel of the first input. The ECU adapts automatically to the throttle input voltage range thereby eliminating the need for throttle adjustments.

A second option is the tri-state throttle switch input which is used to select any one of three precise, pre-programmed, speed commands. For example, in generator set applications, the tri-state throttle can be used to select 1500, 1800 or 2000 r/min engine speeds.

The third basic throttle input is a pulse-width-modulated, PWM, input wherein the throttle command is proportional to the width of the input pulses. This input is used when noise immunity is critical and for throttle commands originating in other intelligent controllers (eg. transmission controllers). A PWM throttle can also be used in conjunction with an analog throttle input.

FUEL TEMPERATURE COMPENSATION - The mass fuel delivery varies with fuel temperature in the inline pump-line-nozzle injection systems. A typical temperature dependence is shown by the dashed line in Figure 9. The engine controller can maintain a constant mass fuel flow (approximately constant power) as the fuel temperature varies over a selected range, eg. the range from $T1$ to $T2$ in Figure 9. This function can also be used to provide power limiting at higher temperatures to reduce engine cooling system requirements. The engines can be rated at somewhat higher power levels when temperature compensation is used since the normal overfueling with cold fuel is reduced. This feature also simplifies power checking of the engines since the power can be held nearly constant over a wide fuel temperature range.

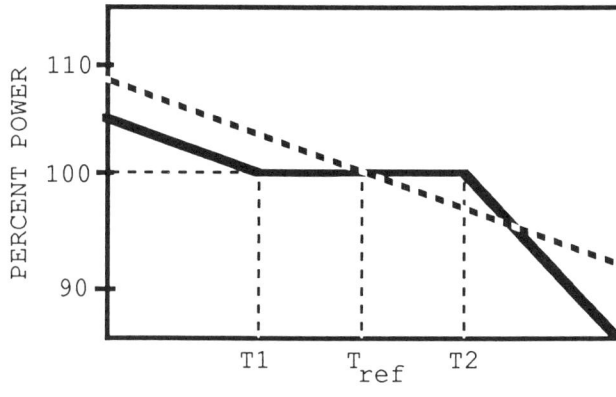

FIG. 9 - Fuel Temperature Compensation

SELF-DIAGNOSTICS AND BACK-UP FEATURES - An engine control system must be highly reliable, but it is also desirable to have a well integrated, on-board, fault detection, warning, back-up and shutdown system. Such a system must be able to accurately diagnose and properly respond if fault conditions occur.

<u>Self-Diagnostics</u> - The Deere control systems include extensive self-diagnostic features. All of the sensor inputs to the controller are monitored for both high and low out-of-range conditions which makes it much easier to isolate the cause of a fault. The critical speed signal inputs are monitored for validity as well as for out-of-range conditions. Missing pulses and noise pulses in the input speed signals are detected and ignored by the controller. The controllers also compare the dual speed sensor inputs with each other to provide an additional level of protection against engine overspeeding.

If a fault condition is detected by the controller, it will both flash a fault lamp and output a digital code over the serial link to notify the operator. These codes are also stored in the ECU and may be retrieved later (even after the ECU has been turned off). Simultaneously, for most fault conditions, the ECU will automatically default to a safe mode of operation. This default mode usually results in nearly normal or slightly derated operation. If a critical failure mode (such as overspeeding with a stuck fuel control rack) is detected, the ECU will respond by automatically shutting down the system. The engine and its installation are protected from potentially damaging operation. Yet the engine is allowed to continue to operate at or near full output if a non-critical fault condition is diagnosed.

Controller Service - Special service connectors are provided for monitoring the fuel flow/throttle commands and the throttle and rack position voltages in addition to the diagnostic codes. The throttle and rack position voltages can be monitored using a standard digital voltmeter. Therefore, in installations which have a code read-out as part of the control panel or dashboard monitoring system, service personnel will normally require only a digital meter to successfully troubleshoot any system faults.

A low cost, hand held, diagnostic reader is available for use as a service tool. This reader can display active diagnostic codes, retrieve stored codes, clear stored codes and display the commanded fuel flow and throttle percentage signals. A list of the diagnostic codes for a control system is given in Table 1.

New electronic systems are typically subject to a high percentage of the initial field complaints resulting in "no trouble found" problems. Thus, in addition to properly diagnosing faults, it is nearly as important for service personnel to be able to verify that the control system is operating properly. The fuel flow and throttle command signals along with the diagnostic codes and voltage outputs provide good means of verifying proper operation of the control systems. Product development engineers have worked very closely with the factory service departments to develop easily followed, yet very complete diagnostic procedures.

TABLE 1 - Controller Diagnostic Codes

Code	Description
0	No Fault Condition Detected
11	Analog Throttle 1 Voltage - High
12	Analog Throttle 1 Voltage - Low
13	Analog Throttle 2 Voltage - High
14	Analog Throttle 2 Voltage - Low
32	Actuator Circuit Fault
34	Rack Position Error
35	Rack Position Voltage - Too High
36	Rack Position Voltage - Too Low
37	Fuel Temperature Voltage - High
38	Fuel Temperature Voltage - Low
39	Primary Speed Error
41	Start Signal Missing
42	Engine Overspeed
43	PWM Throttle Input Error
44	Auxiliary Speed Error
47	Derated Fuel Curve Selected
71	Diagnostic Codes Output - High
72	Diagnostic Codes Output - Low
73	Fuel Flow/Throttle Output - High
74	Fuel Flow/Throttle Output - Low

PROGRAMMABILITY - All application dependent parameters are stored in EPROM or EEPROM memory within the engine controllers. Each controller is programmed at the engine factory to match a specific application at the time the engine is being prepared for final testing (end-of-line programming). The appropriate application data is transmitted into the ECU through the serial communications link. Then the ECU stores this information permanently within its memory. This procedure provides a high degree of control and flexibility for the engine factory while at the same time providing a tamper proof system. A generic ECU and injection pump can be used for a wide variety of applications. Pump resetting is eliminated and parts inventories are significantly reduced. Another benefit to the engine factory is the significant reduction in lead times and costs to introduce new applications.

The controllers can also be reprogrammed if needed to change applications or to fine tune the system for a specific application on site. However, initially such reprogramming will be done only by the engine factory.

ADDITIONAL FEATURES - The control system also includes some additional features such as the key start/stop, a tachometer drive output and the output of the fuel flow and throttle information.

The tachometer drive output is a buffered output of the auxiliary speed sensor signal which is provided by the ECU. This signal can be subsequently monitored by several other systems in the installation without degrading the signal. This same signal is used by the ECU internally as the back-up speed signal.

The controller also provides an output pulse train which contains both fuel flow and throttle information in a special multiplexed format. This output is primarily intended for use by other vehicle control systems. However, this signal can also be used for monitoring fuel flow and for diagnostic purposes.

SYSTEM TESTING & DEVELOPMENT

Special testing equipment and methods were developed to provide thorough testing of these electronic systems prior to production. These included a variety of performance tests, durability tests, environmental tests and production tests.

PERFORMANCE TESTING - A number of tests were developed to verify the functions and actual performance of the engine control systems. These tests included bench tests, engine dynamometer tests, vehicle tests and a number of other special tests such as cold room testing.

The bench tests were used for accurate checking of the operation and functions of the control system, for checking overall system calibrations and for testing software changes.

Engine dynamometer tests were conducted to accurately map the fuel delivery as a function of the rack position voltage and engine speed. Steady-state tests were run to set the torque curves and other parameters such as the initial min-max governor map. Then a series of dynamic tests were performed to optimize the engine response and set the smoke control parameters. Next the governor parameters were usually tuned to optimize the response and stability for a specific application. A series of starting tests under various conditions were used to optimize the starting parameters.

Vehicle or other actual installation tests are conducted to make final adjustments to the controller parameters and verify the overall system performance.

DURABILITY TESTING - The durability of the engine control systems has been proven using accelerated component and subassembly testing, engine life testing and field tests. The bench durability (pump stand) testing is an accelerated cycle test method which has been well correlated with field tests and warranty information. The engine life testing included both steady state, overfueled tests and cycle testing. The in-house component and engine tests have historically proven valuable in accelerating most of the major component failure modes. The field tests in actual vehicles and other installations were useful for developing and verifying wiring practices. The field tests also facilitated the development of proper responses of the control system to intermittent failure modes (some of which are not easily reproduced in the test laboratory).

ENVIRONMENTAL TESTING - Environmental testing of the engine control systems was conducted per rigid in-house testing standards. These tests included vibration and shock tests, temperature-humidity-voltage cycle tests, electromagnetic susceptibility and emissions tests, splash and high-pressure spray wash tests, dust and chemical exposure tests, as well as numerous other electrical tests such as steady-state and transient voltage conditions, open battery operation and short-circuit protection.

Vibration Testing - Vibration testing was conducted based on the MIL-STD-202E with the amplitudes and frequency ranges determined by actual engine or installation vibration measurements at the ECU mounting locations. The systems were powered up and continuously monitored during the entire vibration test cycle to aid in detection of any intermittent failures. Several iterations of controller and actuator designs were required to develop systems which were vibration resistant.

Temperature/Humidity/Voltage Tests - Temperature/humidity/voltage cycling tests were performed with the control systems operating in a special climate chamber. The temperature, humidity and input supply voltage were varied according to a 24-hour cycle, with the total test lasting 100 days.

The governor system was automatically cycled through nineteen test states and continuously monitored during the temperature/humidity/voltage tests by a computerized system. This method of testing was very valuable in identifying potential failure modes which would occur only at temperature extremes and/or under certain input signal and supply voltage conditions. Most of these failure modes are not hard component failures and will not appear during a room temperature functional test.

Electromagnetic Compatibility Testing - Electromagnetic compatibility (EMC) testing was also conducted with the system fully operational and monitored continuously by a computer system since this type of testing does not usually produce hard component failures. Testing of the ECU's and actuators was done in a special anechoic RF-shielded chamber in the lab. An automated test system was used to control the transmitting amplifiers and to monitor field strength and component test voltages. When a sufficient level of confidence was achieved in these component tests, open field testing was conducted with the system operating in a normal vehicle installation.

PRODUCTION TESTING - Special test equipment and procedures were developed for the production testing of these systems. These tests include receiving inspection, quality audit and final engine testing.

Receiving Inspection/ Quality Audit Testing - Portable test equipment is used for complete functional testing of the engine controllers during receiving inspection or quality audit tests. The same equipment is also used for bench performance testing of the injection pumps. Special testing software is programmed into the controllers for these tests to simplify the testing and provide accurate Go - No Go test results.

Factory Final Engine Testing - The same specialized equipment is used to program and run the engine control system in the production factory during the final production engine tests. Again, special software parameters are used to facilitate these tests.

APPLICATIONS

The engine control systems are designed to be flexible so they can be readily adapted for use in a wide variety of applications. A number of these applications were included in the development testing. The engine controls are in production for generator sets and various agricultural vehicle applications. Various other off-highway applications are scheduled for production.

GENERATOR SETS - A very cost effective OEM application is in generator sets. The electronic engine control system can replace the add-on electric governors for lower cost and provide better isochronous governing and a number of other benefits such as the precise switch selectable speeds using the tri-state throttle which is immune to drift and requires no adjustments. The programmable governing parameters facilitate the matching of the engine governing to the generator set applications. The power boost feature could be used to provide a higher power limit for a short time to aid in handling short duration loading during start-up of electric motors. A derated operating mode can be selected by user supplied switch inputs (eg. temperature or pressure) to provide an intermediate level of engine protection without having to completely shutdown.

The back-up speed sensor and the ECU control of both the actuator drive and the fuel shutoff valve provides a redundant level of protection against engine overspeeding.

OFF-HIGHWAY APPLICATIONS - The capability for readily tuning the control system for optimal engine and vehicle performance is a significant advantage of electronic controls. The torque curves, governing characteristics, engine response, engine starting, etc. can all be matched to each application. The elimination of the throttle and shutoff linkages gives new flexibility to the vehicle designers.

Several unique applications of the control system features are found in the off-highway vehicles. The power select (tri-state input) is used to provide a power boost mode on combines for unloading the grain tank "on the go". Power deration is selected by a temperature switch to protect the engine if the air-to-air intercooler plugs. The power select feature is also used to limit the maximum engine torque to protect the drive train yet allow for extra power when needed for hydraulic pumps, etc.

The isochronous low idle speed regulation improves the load starting ability and helps in maneuvering vehicles at low speeds. The speed regulation and rated speeds can be selected by the tri-state input to limit the engine speed for power-take-off operations or to increase the maximum engine speed for transport. The tri-state throttle is used in off-highway vehicles which have hydrostatic drives to provide a simple throttle input.

ON-HIGHWAY APPLICATIONS - Many of the features used in off-highway applications are also very useful in on-highway applications. These include the isochronous low idle speed control, the precise torque curve shaping, key ON-OFF, torque limiting to protect the engine or drivetrain, self-calibrating throttle inputs, smoke control and starting improvements. Other features such as min-max governing and road speed limiting were included primarily for on-highway applications. The fuel flow/throttle output and the PWM throttle input signals enable these systems to be fully integrated with other vehicle systems such as transmission controls.

SUMMARY OF DEERE ENGINE CONTROLS

The new Deere engine control systems provide full-featured controls for use in a wide variety of applications. These systems are designed to be used in heavy duty, rugged applications while at the same time providing precise engine control. Significant development work has been done to provide very cost effective, highly reliable engine controls. The flexibility of these electronic controls gives them a major advantage over mechanical controls in adapting to new application requirements.

890392

Control Design for a Differential Compound Engine

J. Hall and F.J. Wallace
School of Mechanical Engineering
University of Bath
United Kingdom

ABSTRACT

A microcomputer-based controller has been designed for a differential compound engine (DCE). Following assessment of the DCE state interactions, single input-output loop strategies were developed using a dynamic simulation. The simulation, control strategies and final implementation are described, and simulated transient responses are presented.

THE DIFFERENTIAL COMPOUND ENGINE (DCE) is an integrated engine-transmission, its main design principles being the connection of diesel engine, compressor and load through a torque-balancing epicyclic gear to provide a continuously variable transmission ratio, and the use of an engine air bypass and compounding turbine to provide substantial torque back-up.

From the control engineer's viewpoint, the DCE poses the same challenges as other power-trains incorporating electronically-controlled transmissions. With a single feedforward "demand" from the driver, the controller must generate engine fuelling and transmission ratio control inputs to achieve optimum fuel efficiency and fast, stable, transient response within a range of legislative and engineering limits.

This paper describes the design process leading to the implementation of a micro-computer-based controller in a prototype DCE.

DCE CHARACTERISTICS

Mechanical description - the design and component matching of the laboratory prototype on which this work was conducted was reported in (1). The DCE is shown schematically in Fig.1; the major components are :

Engine :   Leyland 500 series 8.2 ltr.6 cyl
              DI Diesel
              max. BMEP 21 bar)   in this
              max. power 234 kW)   application

Compressor : CompAir rotary positive
              displacement (dry screw type)
              max. speed 11500 rev/min.
              max. pressure ratio 4.0

Power Turbine : modified Napier C045 radial
              inflow with variable geometry
              nozzles.
              max. speed 50,000 rev/min.
              max. pressure ratio 4.0

Gearbox :   Allen Gears combined epicyclic/
              layshaft gear train.

Relationships of DCE state variables - the DCE gearing arrangement and components impose relationships between the parameters which describe the state (operating condition) of the DCE. The epicyclic gearing arrangement gives the following characteristics :

(i) engine, compressor and output shaft torques are in fixed ratio at steady-state, irrespective of speeds.
(ii) the speed ratios are not fixed; any two speeds may be set independently, the third is then dependent.

The rotary positive displacement compressor gives the additional properties that
(iii) boost pressure ratio is approximately proportional to the epicyclic torques.
(iv) system massflow is approximately

* Numbers in parentheses designate references at end of paper.

proportional to compressor speed.

The above properties lead to the conclusion that the fundamental state of the DCE can be fully described by three independent parameters, viz. [any epicyclic torque, OR boost pressure ratio] AND [any TWO epicyclic-connected component speeds (engine, compressor or output shaft) OR system massflow (in lieu of compressor speed) and another epicyclic component speed].

DCE CONTROLS

The prototype DCE has four control inputs: engine rack, turbine V.G nozzles, and fuel pump timing, and output shaft load/speed control.

Rack control - A Sigma in-line fuel pump is fitted, with mechanical governing removed. The rack is hydraulically actuated via a Moog servovalve. Position feedback is by a.c. LVDT with remote conditioning. An analog control loop gives positional control. The servo control will accept manual or computer feed-forward inputs, and can stand alone or be incorporated as a basic inner loop for other controls. Rack, nozzle and timing control all use a similar arrangement (fig. 2).

The fuel injection equipment (FIE) is thus in the class of first-generation electronic FIE, employing a conventional fuel pump with mechanically set fuelling and injection timing, rather than pulse modulated injection.

Nozzle control - The turbine variable geometry arrangement uses swivelling profiled vanes. Each vane is actuated by an individual arm clamped to its shaft. The arms are attached to a unison ring floating on roller bearings. The unison ring is hydraulically actuated. The nozzle shafts use a Stellite 31 coating ground to give a free running fit in Stellite H66 bushes at the "worst case" thermal expansion. The arms are driven from the unison ring by a peg and slot arrangement. At present, lubrication is by a single shot of an aluminium-based high temperature anti-seize compound during assembly. Insufficient test hours have been run to safely claim good durability, but no problems have been experienced.

The range of nozzle movement is from nominally closed (flow due to edge leakage only) through 30°, as shown in figs. 3 and 4 respectively. This gives a 9:1 turndown ratio in choked swallowing capacity.

Time control - The fuel pump is driven via a helical spline, providing a static injection timing range of 22° crank.

Load (dynamometer) - Output shaft loading (constant torque or speed) is by twin axial piston pumps.

Effect of controls on the DCE state - As explained above, the basic DCE state is fully defined by three independent variables. Neglecting the relatively minor effect of injection timing on the operating condition, there are three primary controls. Two are available to the controller (rack and nozzle position), the third (output shaft speed/load) is set by the dynamometer, or in a practical application by the vehicle and road conditions. There is no direct correspondence between the control inputs and three independent state variables. With the output shaft in general only loosely controlled by the load conditions, the rack and nozzle controls will both affect component speeds and the epicyclic torque (and thus boost) level.

In summary then, the DCE is a multivariable system, where the two available primary control inputs (rack and nozzle position) can both affect any state variable.

SIMULATION

Given a non-linear, non-stationary (response changes with operating condition) system such as the DCE, three practical approaches to controller design are possible :
(i) Identify and linearise the response of each controlled parameter to each control input to obtain a state-space or transfer function model. Design compensators using established techniques such as pole placement or Bode plots.
(ii) Select the form of the control loops qualitatively, and tune and test the controller systematically using a simulation package.
(iii) Establish the approximate form of the response, and carry out identification and compensator design for each control loop in real time, i.e. adaptive control.

Ref. 2 describes the application of approach (i) to an automotive gas turbine. The two-shaft gas turbine is closely analogous to the DCE in that the control inputs (gas generator fuelling and free turbine nozzle angle) have strongly cross-coupled effects on the main operating parameters. Having carried out identification and linearisation, the authors of Ref.2 were able to design multivariable precompensators using Rosenbrock's inverse Nyquist array technique (Ref. 3). However, the individual loop compensator terms were chosen qualitatively and tuned systematically using a dynamic simulation.

It was therefore clear that a simulation program would be required for the DCE controller design, whichever approach were to be used.

Specification - For controller design, the simulation should model system dynamics accurately but need not have great thermodynamic accuracy. To be a useful design tool, it must have short run times. In this case, the software was implemented on a DEC LSI11/23 microcomputer, coded in FORTRAN.

Structure - The structure of the simulation ('SIMDCE') is shown in fig. 5. The most important subroutines are EULER1 - a numerical integration routine using the predictor-corrector method, DCECSZ - the DCE model, and CTRLR - the controller model. The inputs to the program are a single feedforward demand and the load condition; this may be either a load torque or a vehicle/road condition which is translated to the equivalent driveshaft load torque at each timestep by subroutine VEHICL.

DCE model - The simplicity (and thus speed) of the basic model is based on the ability of three independent variables to define the DCE state. By treating all manifolds and plenums as a single control volume, only three dynamic terms (chosen to be output shaft acceleration, compressor acceleration and rate of change of boost pressure in the control volume) need be carried along by the numerical integration routine. The acceleration dynamics are based on accurately-known component inertias. The pressure dynamics are based on known pipework volumes, compressor mass-flows and turbine swallowing capacities, but loss factors were introduced to match results to a more comprehensive simulation which models pipe friction and heat losses.

All non-dynamic calculations use empirical arrays or linearised relationships based on experimental steady-state data from the prototype, and are thus quasi-steady. However, note that since boost pressure is modelled dynamically such factors as its effect on engine air/fuel ratio and thus efficiency and exhaust temperature are modelled more accurately than in a completely quasi-steady simulation. In addition, the effect of discrete injection has been modelled by a first-order lag in torque response, as suggested in Ref.4.

Actuator models - Theoretical calculation of the closed-loop responses of the electrohydraulic actuators predicted a third-order transfer function having (for the rack actuator) near flat response up to 20 Hz, with 90° phase lag at 110 Hz. For comparison, the engine firing frequency at rated speed is 130 Hz. On this basis it was not clear whether the computational overhead of extra dynamic states for the actuators was worthwhile. However, on identification of the rack and nozzle actuators, the responses were found to be much poorer than predicted. In terms of the 90° phase lag frequency :

    Theoretical rack (closed loop)   110 Hz
    Measured rack (closed loop)   6.5 Hz
    Measured nozzles (closed loop)   16 Hz

The significant rack phase lag at these low frequencies is probably due to the sliding friction of close-tolerance fuel pump components. Both actuator responses were therefore included in the DCE model. Second-order parametric models were used, obtained by an interactive program plotting a range of models onto a Nyquist diagram, compared to the identified response. Each second-order equation was reduced to a pair of first order equations (by definition of a new state) which could be numerically integrated along with the three main DCE state variables.

Controller subroutine - The controller subroutine is separate from the DCE model, since it will be continually modified to simulate alternative controller designs. In all cases the controller routine will read the "demand" feedforward, and generate control inputs according to the current control design and the operating condition.

The model and controller may be run with differing timesteps, to simulate the effect of a digital controller's discrete time, rather than continuous, operation.

CONTROLLER DESIGN

This first phase of work, designs based on single input-output (SIO) loops were considered. The rack input is used to control engine speed, and the feedforward demand is taken to be engine speed demand. Direct connection between the feedforward and a single control input simplifies the design; the nozzle control input thus has the following tasks :

Steady-state - With engine speed set by the feed-forward demand, and torque level set by the (extraneous) load, the nozzles must set the third independent state variable (chosen to be compressor speed) to achieve best thermal efficiency. Open-loop nozzle position scheduling was ruled out because of potential V.G mechanism unrepeatability.

It is not necessary to schedule compressor speed according to engine speed and load torque; any two convenient parameters may be used provided they and compressor speed are a set of three independent state variables.

Transient - During a transient the epicyclic torque and boost relationships no longer hold. The nozzle control may thus set any boost pressure irrespective of engine torque. Furthermore, the transient objective is to achieve maximum output shaft acceleration or load acceptance within the engineering boost pressure limit (4.0 bar in this prototype).

Results are presented for two finalised designs. Control terms were introduced and tuned as necessary to achieve tight, stable control for a range of simulated conditions.

Oscillatory behaviour was experienced as a result of

(i) coupling between engine speed and compressor speed control loops. Nozzle angle controls system swallowing capacity and thus alters boost level, causing the compressor speed to change as its load varies. However, this also changes engine loading, giving an unwanted 'noise' feedpath into the engine speed control loop. In addition, since the scheduled compressor speeds are mapped on a base of rack position versus output shaft speed (fig. 6) changes in rack position cause the desired compressor speed to change whilst the controller is trying to set it.

(ii) transitions between steady-state and transient operation, when the nozzle control input must change between compressor speed control and response/boost control.

Controller C1 - Fig. 7 shows the controller structure, table 1 summarises the control terms used, as discussed below:

PARAMETER	CONTROL INPUT	CONTROL TERMS
engine speed	rack	P
compressor speed	nozzles	PI
boost/fuelling ratio	nozzles	PD
"	rack	variable max. limit
maximum boost	nozzles	PD
injection timing	timing	positional only

P = proportional
I = integral
D = derivative

TABLE 1

The compressor compensation incorporates an integral term to drive steady-state errors to zero. Boost/fuelling ratio is used in lieu of engine air/fuel ratio for exhaust smoke control. The nozzles hold a minimum boost pressure level to keep within exhaust smoke limits at the current fuelling. Since engine fuelling response dynamics are at least an order faster than boost dynamics, initial transient overfuelling would still occur, despite the use of a derivative term to 'anticipate' low boost/fuelling ratios. A conventional boost-related fuelling limit was therefore added. Note that the use of the fuelling limit alone would penalise transient response unnecessarily since the DCE transient boost can be directly controlled, c.f. conventional turbocharged engines where boost response is predetermined by turbocharger match and inertia.

To avoid introducing additional coupling between rack and nozzle control loops, the required boost level is based on the desired fuelling rather than the boost-limited fuelling.

Controller C2 - The large number of control terms in C1 required considerable tuning. C2 uses a simplified form of the above strategy, requiring fewer terms (fig. 8 and table 2).

Integral control of compressor speed was replaced by scheduled nozzle position offsets. Since the offsets were high (typically 80-95% closure), the proportional term could be reduced, improving stability and smooth transition from transient control. Transient (i.e. boost/fuelling ratio) control is switched discretely. The most effective transient measure was considered to be rack position. The switchover is made at a predetermined level, in this case maximum rack. Since the transient control is only invoked at maximum rack, the required boost level is a fixed value.

Predicted transient responses - Results are presented for C1, C2, and C2 with transient strategy disabled (C3). Three transients are shown, all with the dynamometer inertia only:
  i) Demand step : 50% to 100% demand at 500 Nm load.
  ii) Load step   400 to 1400 Nm load at 100% demand
  iii) Combined step : 45% demand, 800 Nm load to 90% demand, 1600 Nm load.

Figs. 9a,b,c show demand step responses for C1, C2 and C3 respectively. C2 has faster boost and output shaft speed response, though the discrete switch from transient to steady-state control at about 7.5 sec. causes a kick in boost pressure. Load acceptances (figs. 10a,b,c) are very similar due to the inherent DCE characteristic of increasing compressor speed (and thus boost) with a stalling output shaft. Note the oscillatory nozzle response of C1 due to generally higher control gains. Again in the combined step (figs. 11a,b,c) the oscillatory nozzle control of C1 gives poor boost response and a greater initial dip in output shaft speed (in each case smoke-limiting control restricts fuelling and thus torque with inadequate boost). C3 also shows a poorer response due to slow boost build-up, demonstrating the need for differing steady-state and transient strategies.

Design C2 was chosen for implementation on the prototype.

Discrete control rate - Before implementation on a microcomputer-based controller, discrete control operation was simulated by running the controller subroutine asynchonously from the model with the analog loops discretised. Ref. 5 describes practical discretisation rules; in this case simple backward-difference approximations were preferred.

Fig. 12 shows load step responses for controller C2 operating continuously and at two (low) frequencies. Please note the lower analog case initial torque (due to the loose simulation initial convergence limits used). A target controller rate of 200 Hz was selected.

CONTROLLER IMPLEMENTATION

Controller C2 was implemented on a Dell 200 (80286 AT-compatible) micro, with the off-the-shelf PC-bus 12 bit ADC/DAC boards. The code was written entirely in C, using long integer and

PARAMETER	CONTROL INPUT	CONTROL TERMS
engine speed	rack	P
compressor speed	nozzles	offset + P
boost/fuelling ratio	nozzles	offset + P, <u>fixed</u> required boost level
injection timing	timing	positional only

TABLE 2

integer (16 bit) types. Interactive bumpless transfer between manual and micro control was provided, with signal error checking and emergency idle/shutdown facilities. Fig. 13 shows the hardware layout. The controller runs at 580 Hz. Steady-state trials have been completed. The DCE behaves stably at all conditions. Slight changes to the timing schedule on the limiting torque curve were required to achieve the best max. cylinder pressure/exhaust temperature trade-off. Steady-state compressor speeds were held acceptably close to scheduled values, although more detailed scheduling may be required (currently 56 points) since optimum compressor speed changes quickly at high fuellings.

Transient test instrumentation is currently being commissioned.

CONCLUSIONS

A simulation package specifically designed for :
  (i) accurate modelling of dynamics
  (ii) fast execution by the use of quasi-static empirical data for non-dynamic calculations
has been used to develop conventional single input-output controller designs for a complex multivariable powertrain, leading to the implementation of a microcomputer-based controller. The simulation will also be required to validate and develop controllers using real-time identification/control-law design, planned as the next phase of the DCE project.

ACKNOWLEDGEMENTS

The above work was part of a major DCE research project funded by the U.K. Science and Engineering Research Council.

REFERENCES

(1) Wallace, F.J. et al
    "Design and Performance Characteristics of the Laboratory Differential Compound Engine at Bath University", I.Mech.E.C196/86.

(2) Winterbone, D.E. et al
    "A Multivariable Controller for an Automotive Gas Turbine", ASME Conf. 1979.

(3) Rosenbrock, H.H.
    "Computer-Aided Control System Design", Academic Press 1974.

(4) Hazel, P.A. and Flower J.O.
    "Sampled Data Theory Applied to the Modelling and Control Analysis of Compression Ignition Engines", Int.Journal of Control, Vol 143, No.3, 1971.

(5) Katz, P.
    "Digital Control using Microprocessors", Prentice-Hall, 1981.

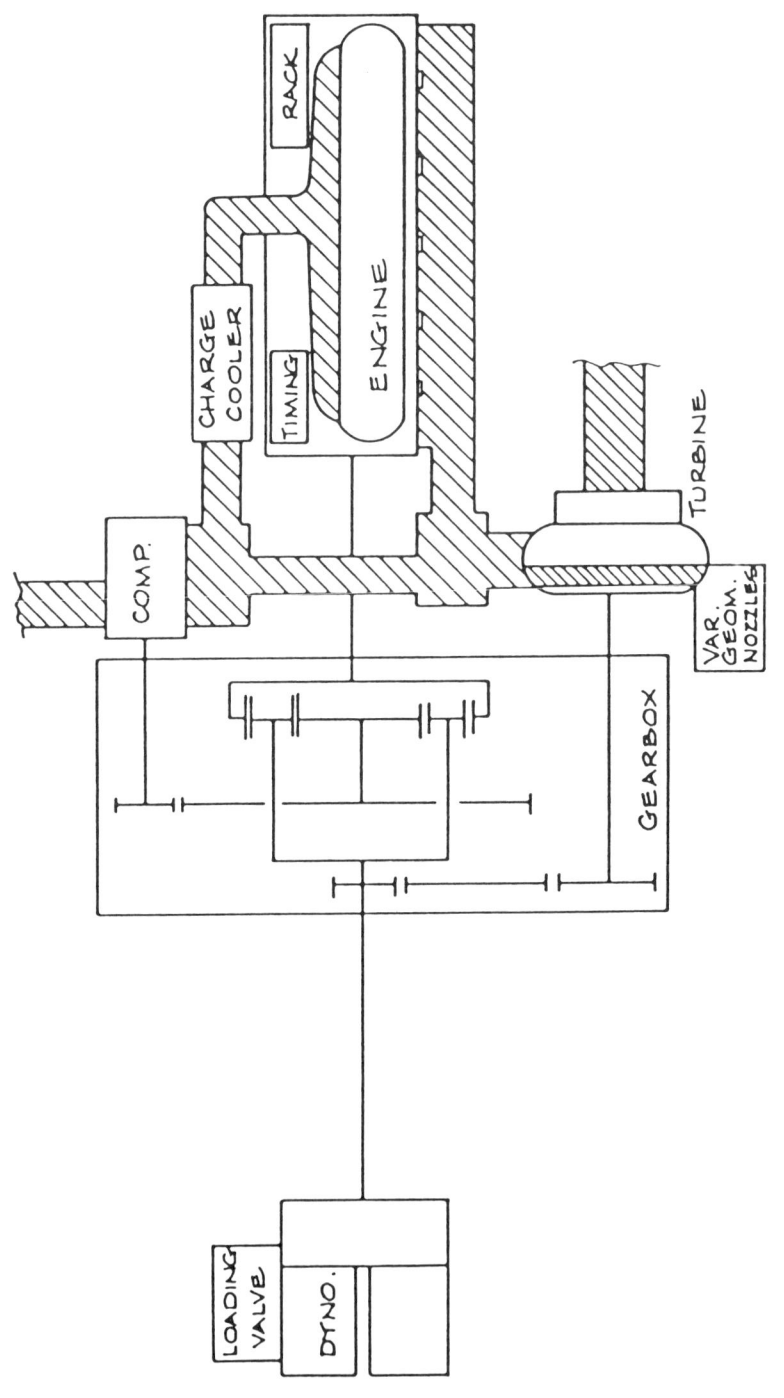

FIG.1

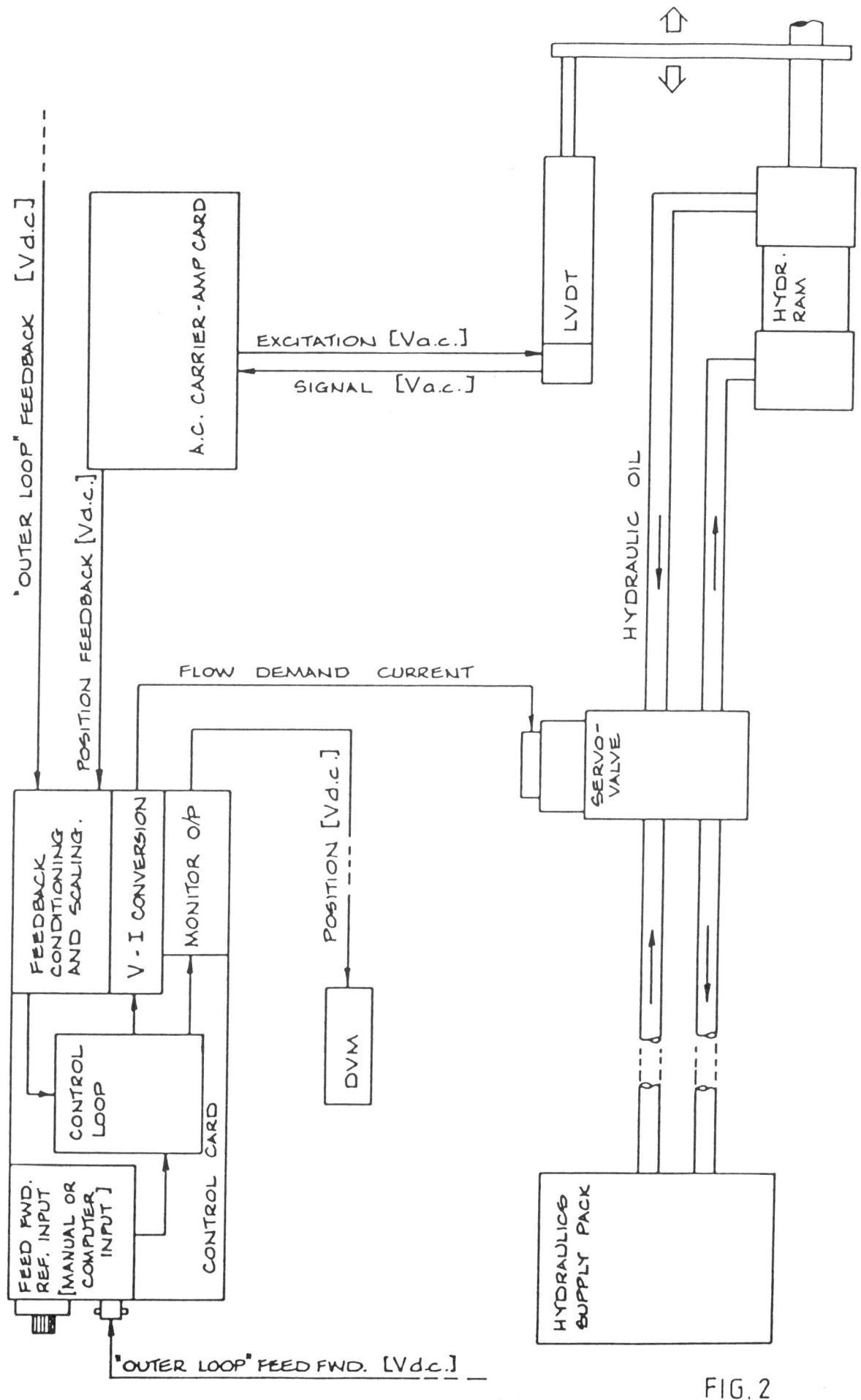

FIG. 2

FIG. 4

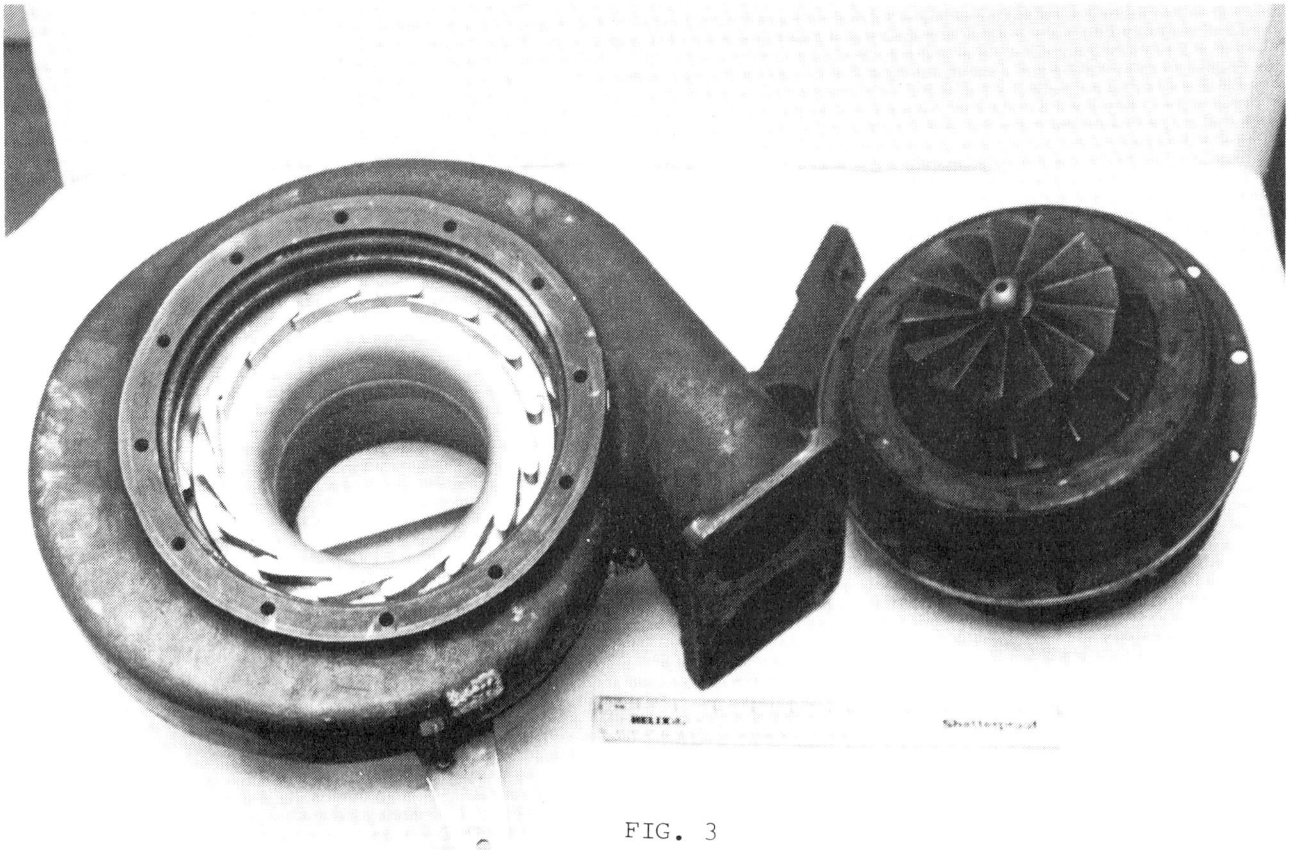

FIG. 3

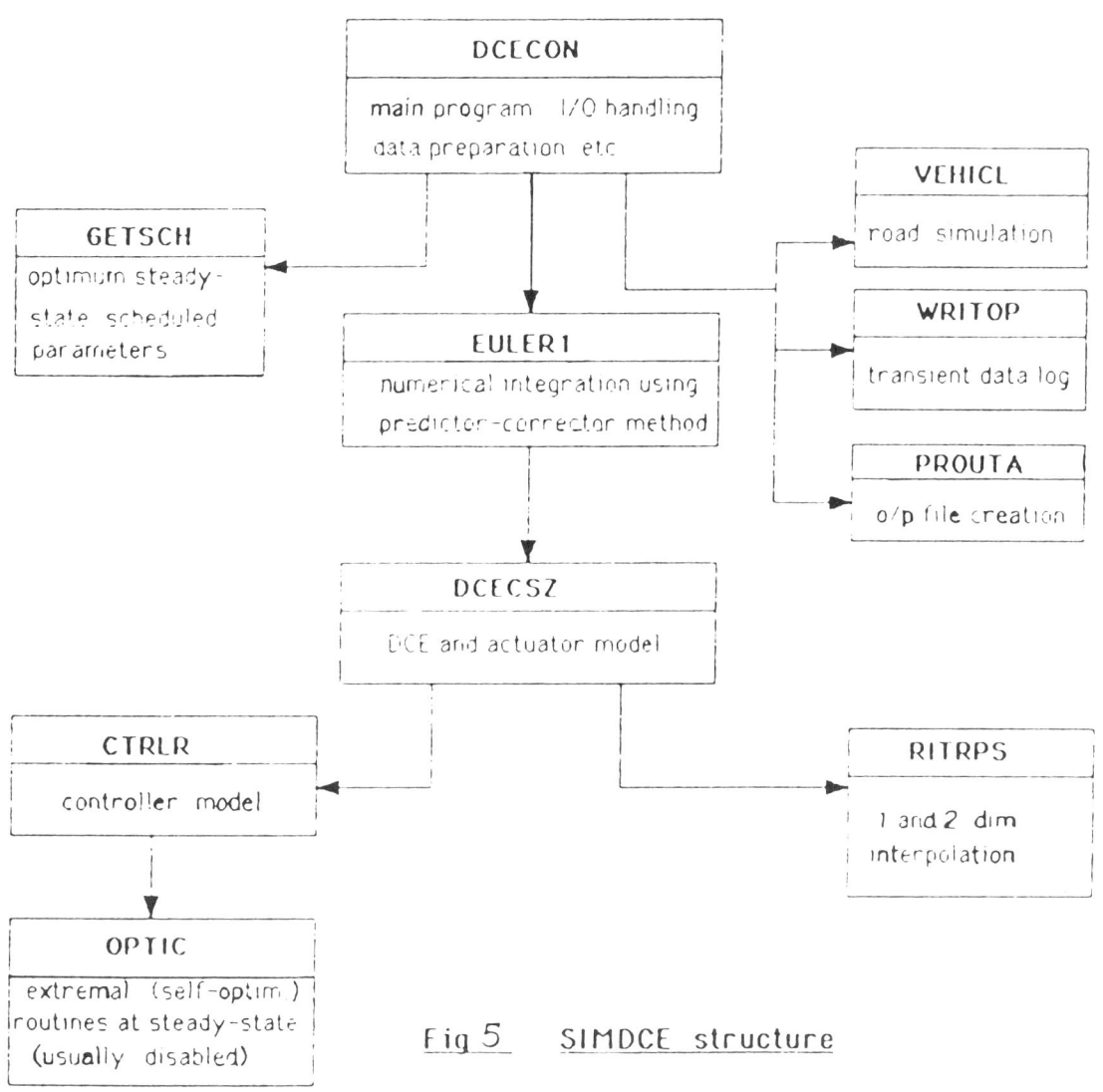

Fig 5   SIMDCE structure

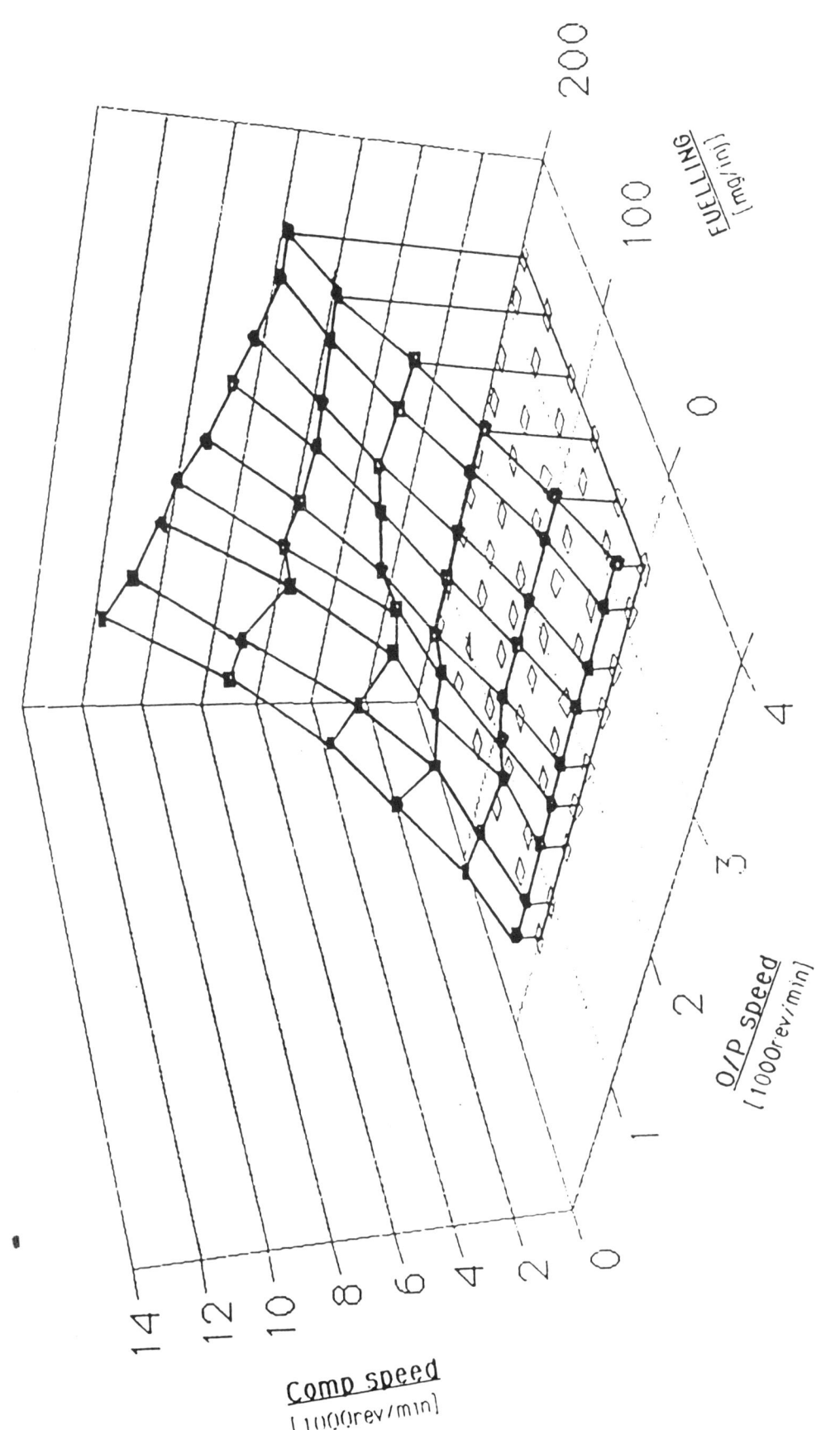

FIG. 6 OPTIMUM COMPRESSOR SPEEDS

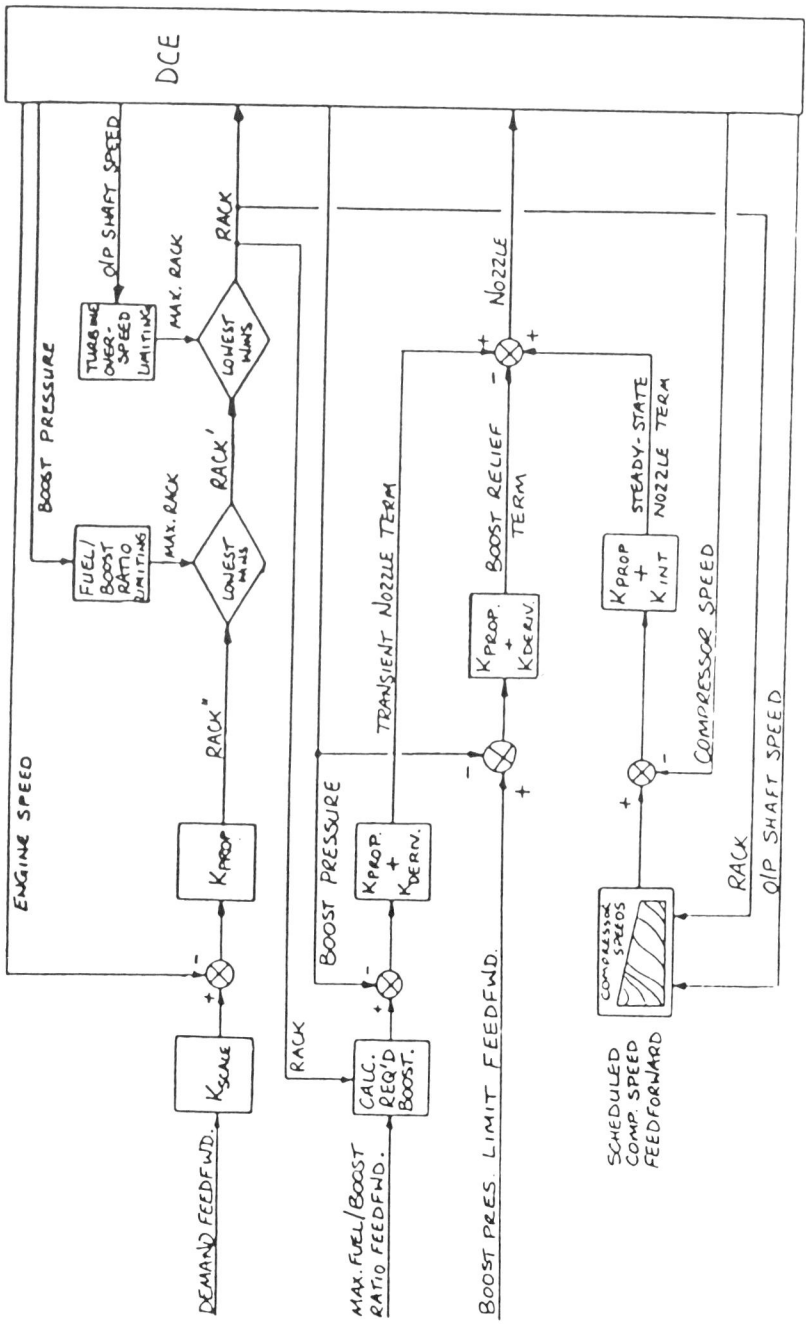

FIG. 7   CONTROLLER SCHEMATIC – C1

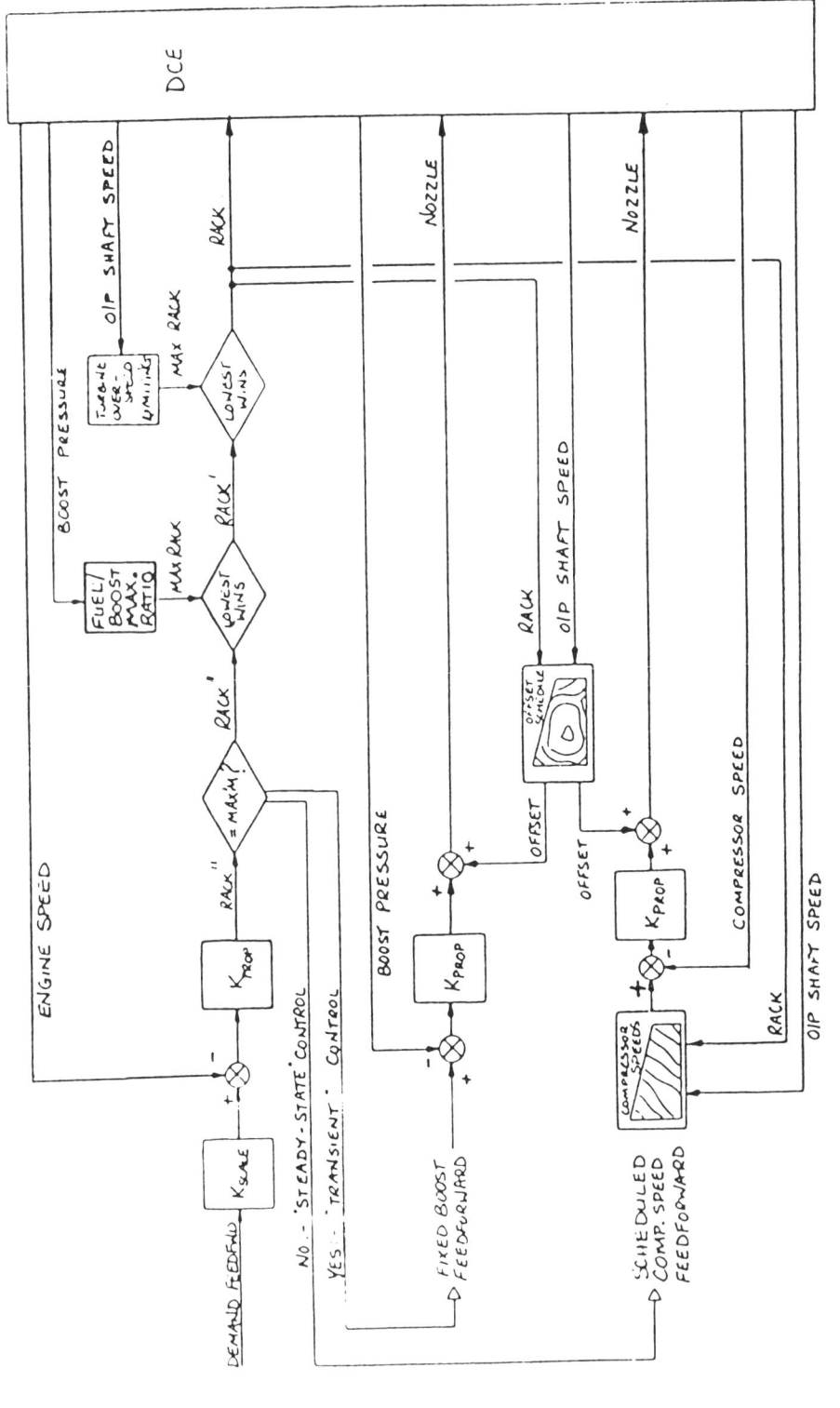

FIG.8　CONTROLLER SCHEMATIC - C2

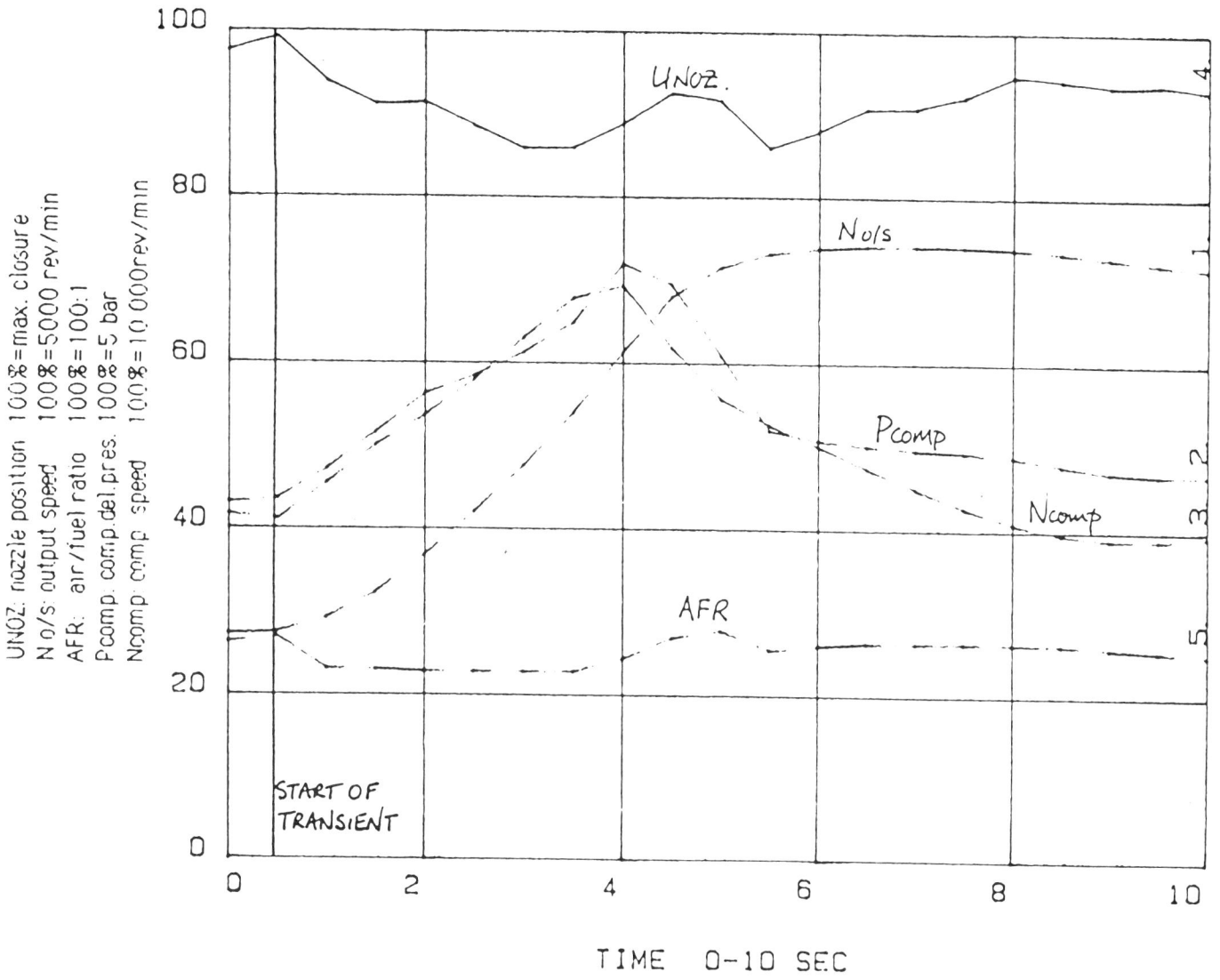

Fig 9a   Demand step: C1

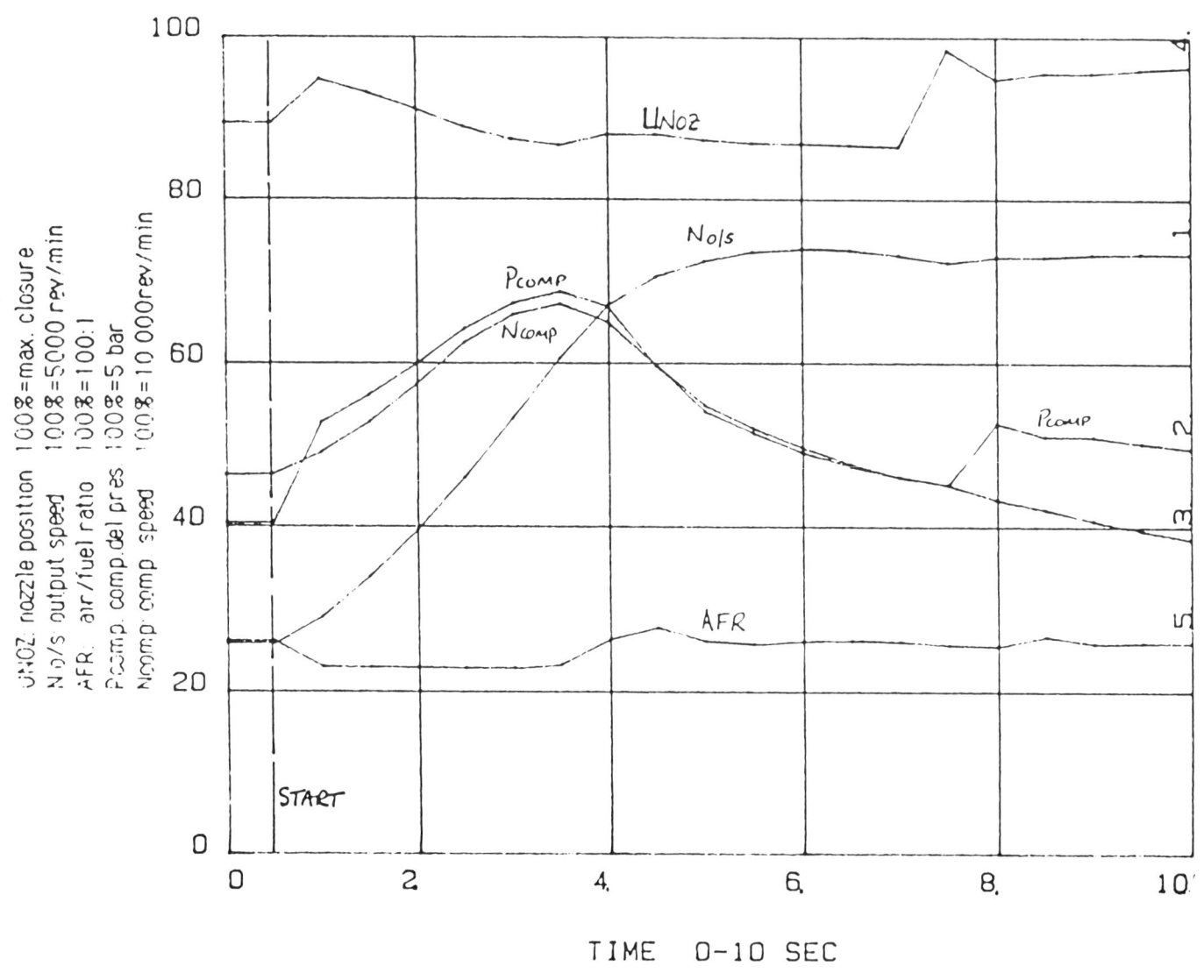

Fig 9b   Demand step: C2

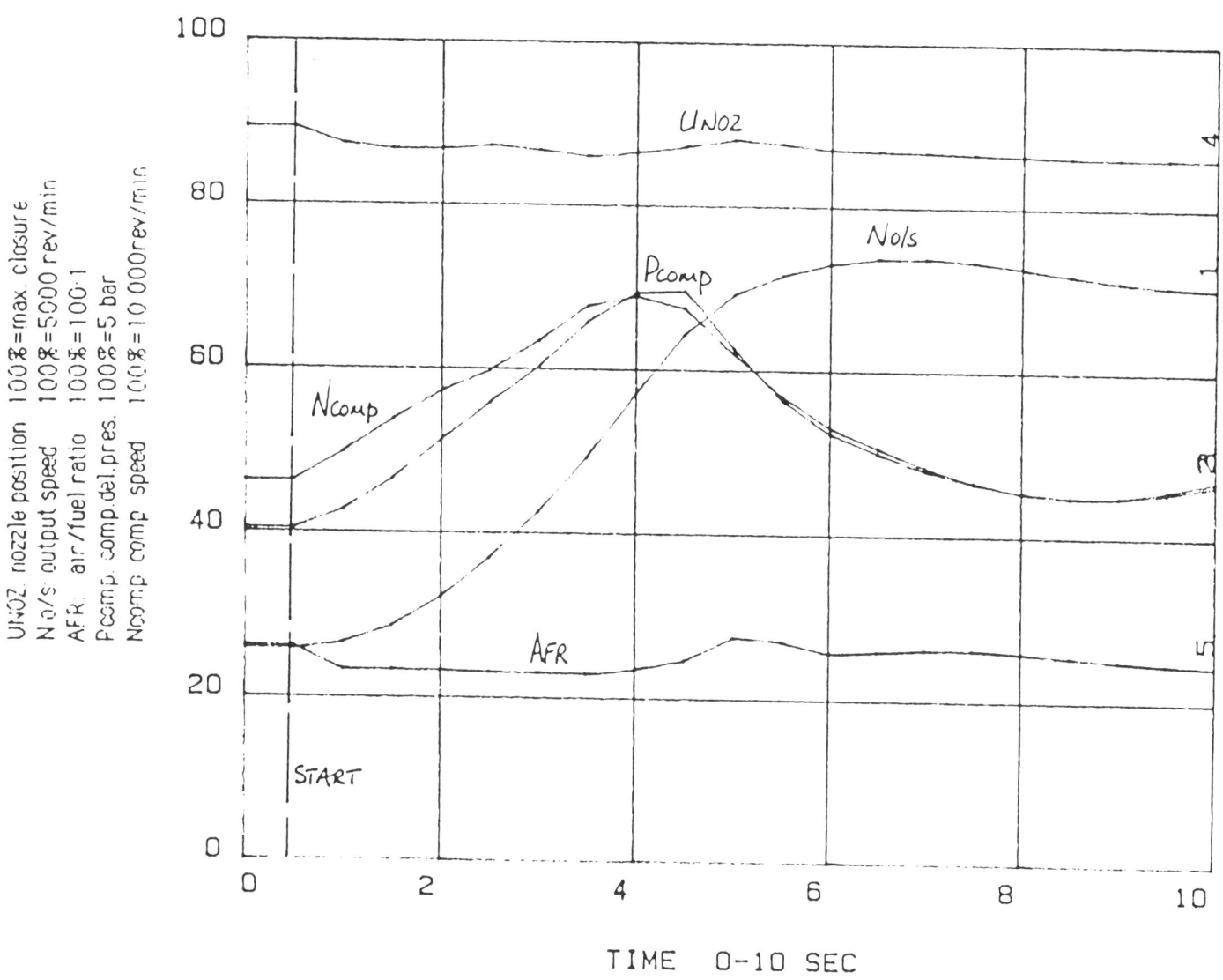

Fig 9c   Demand step: C3

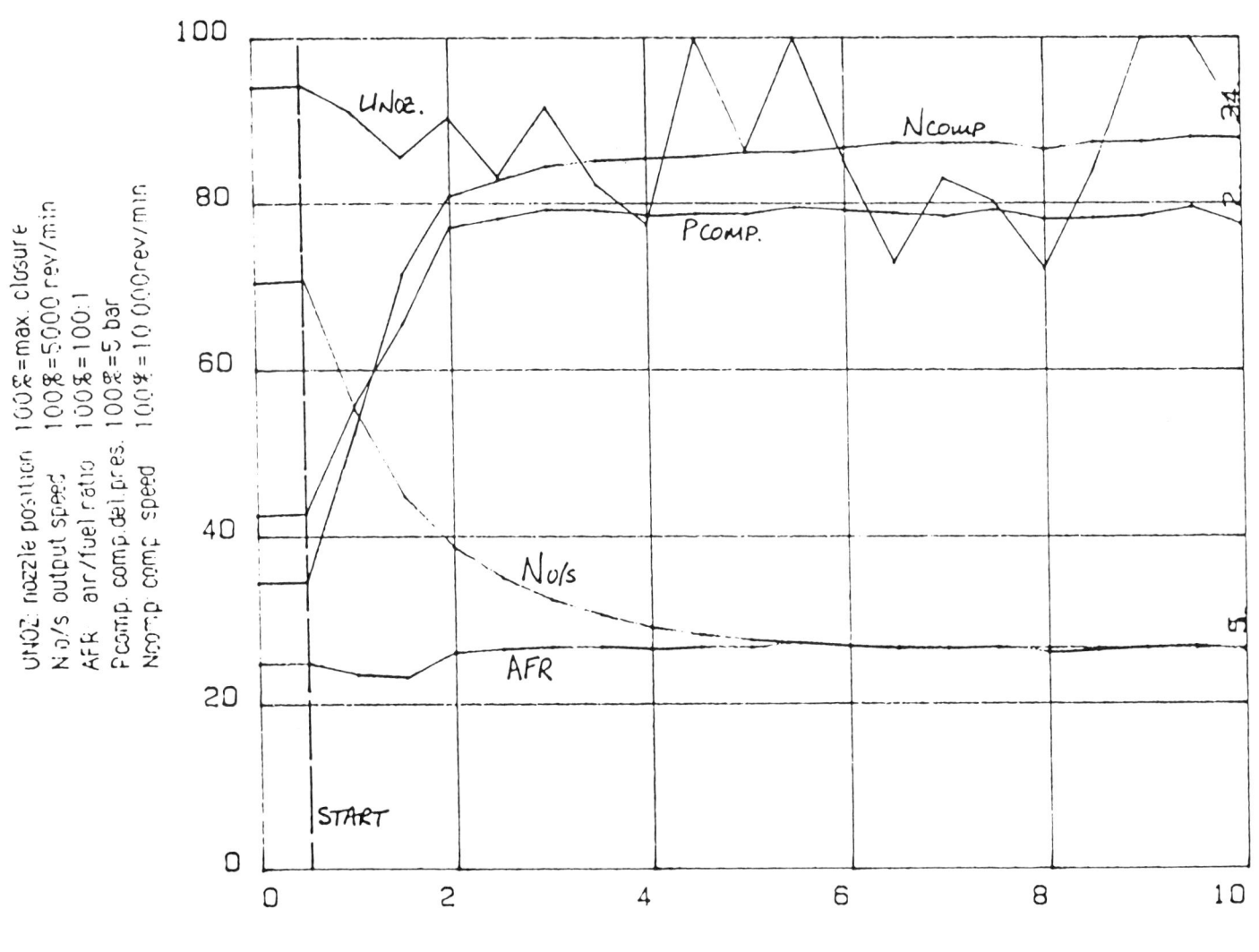

Fig 10a   Load step: C1

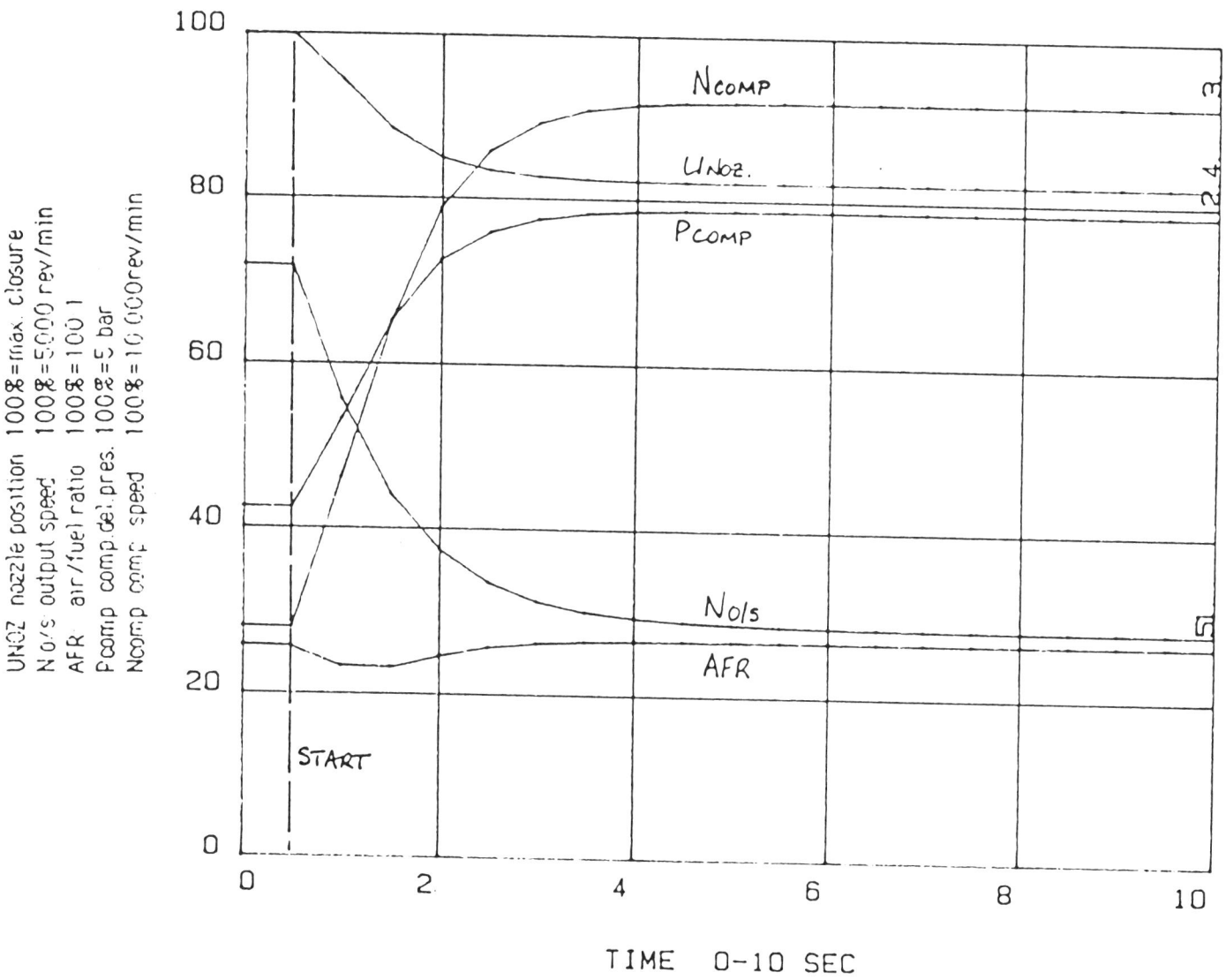

Fig 10b Load step: C2

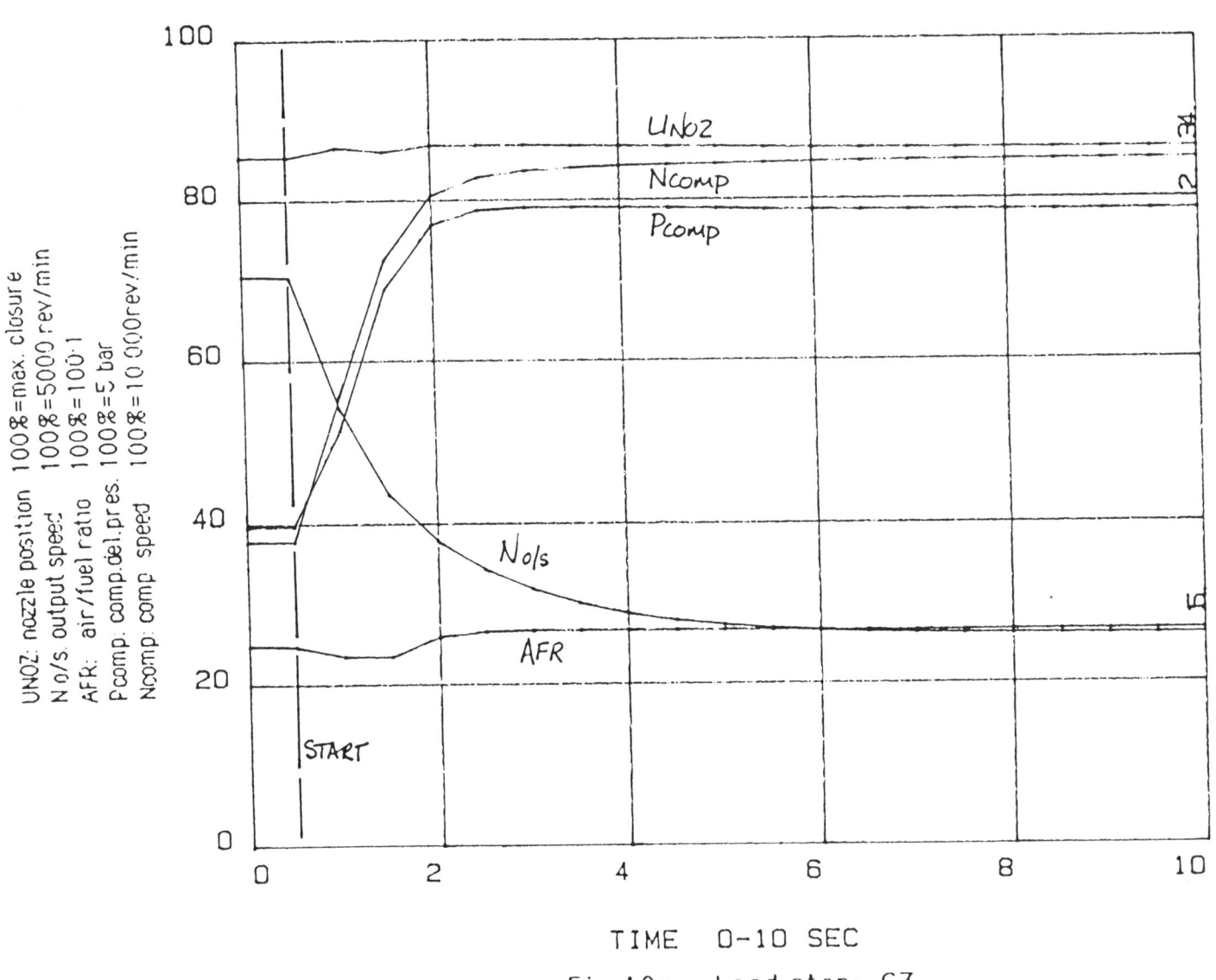

Fig 10c   Load step: C3

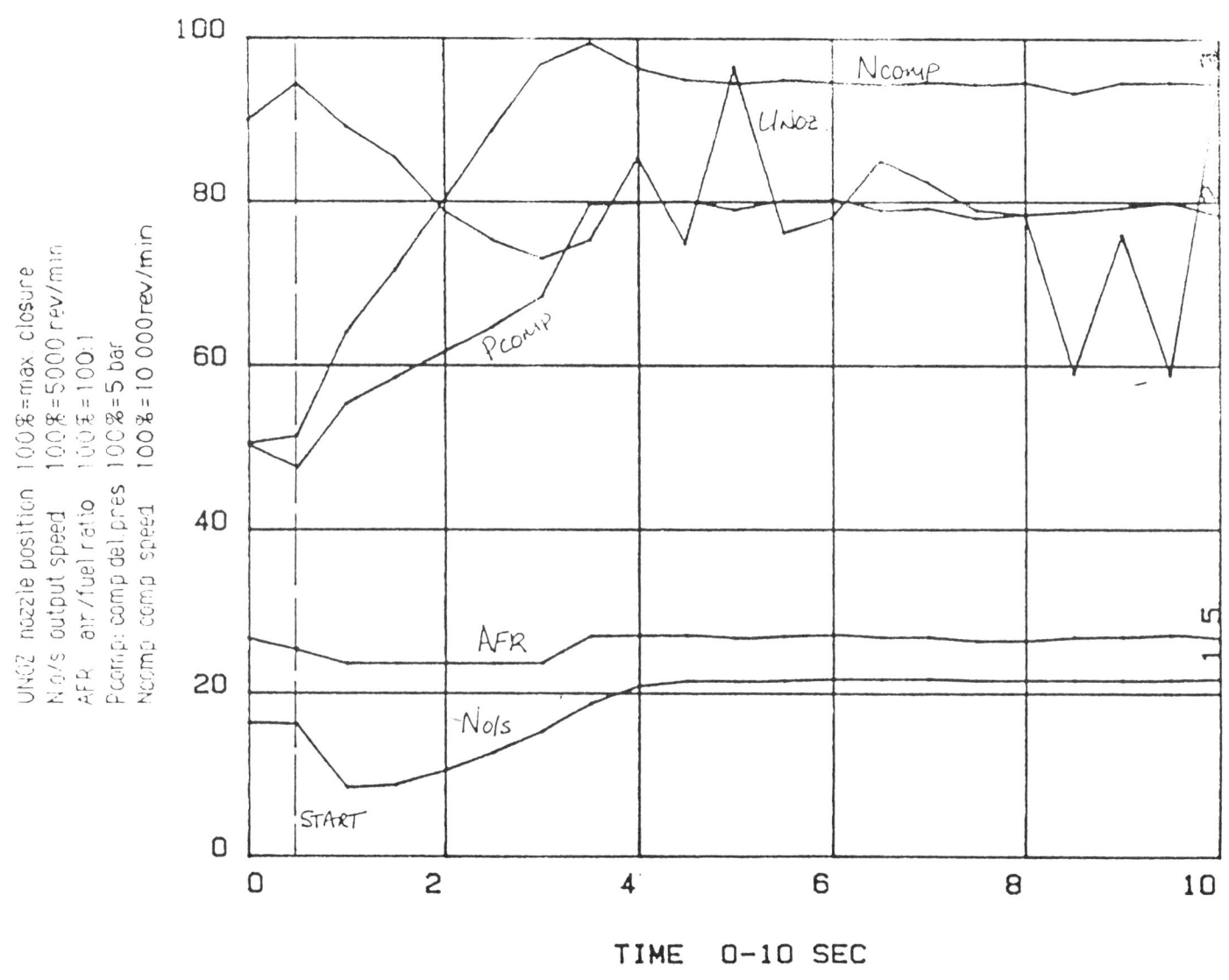

Fig 11a Combined step: C1

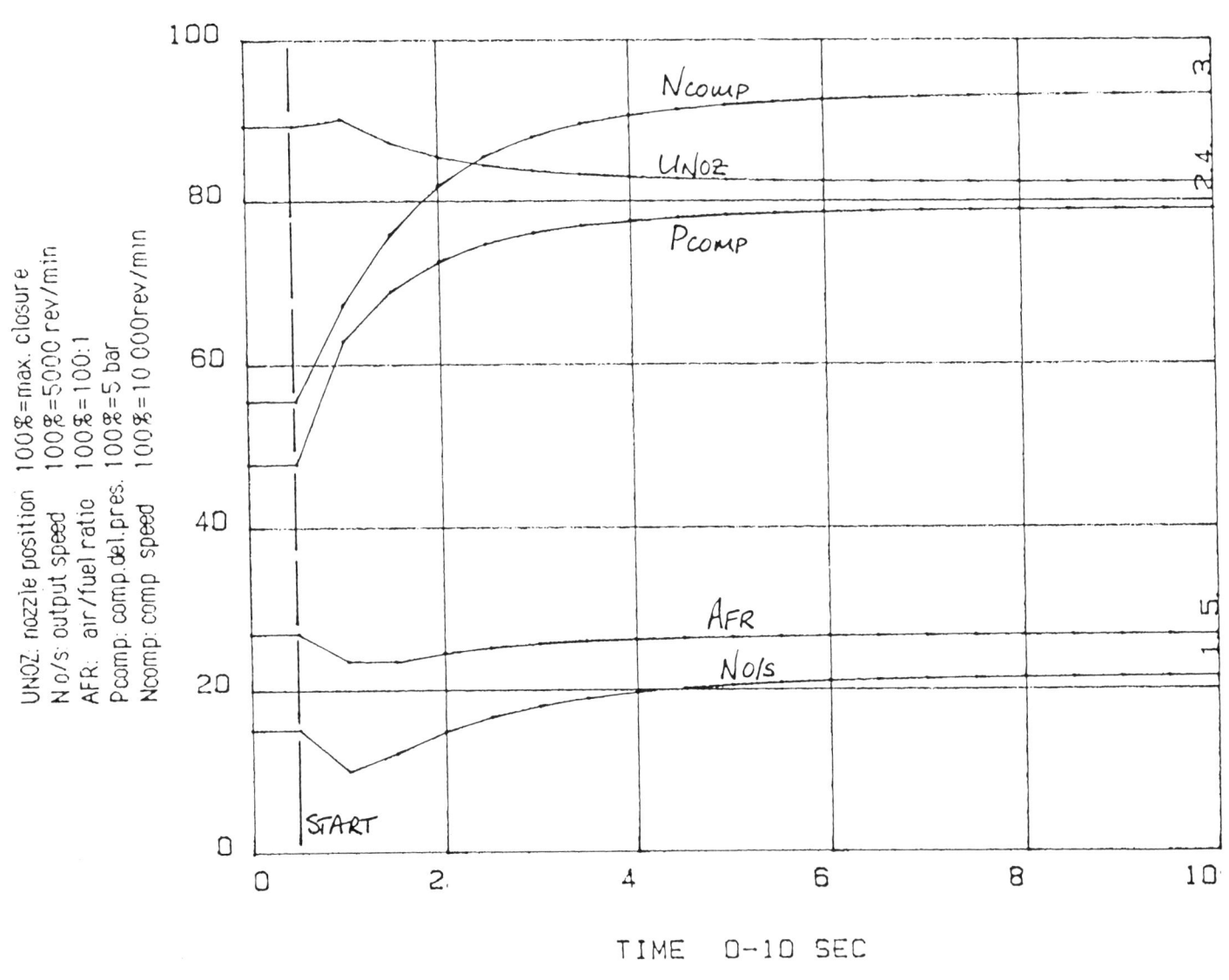

Fig 11b Combined step: C2

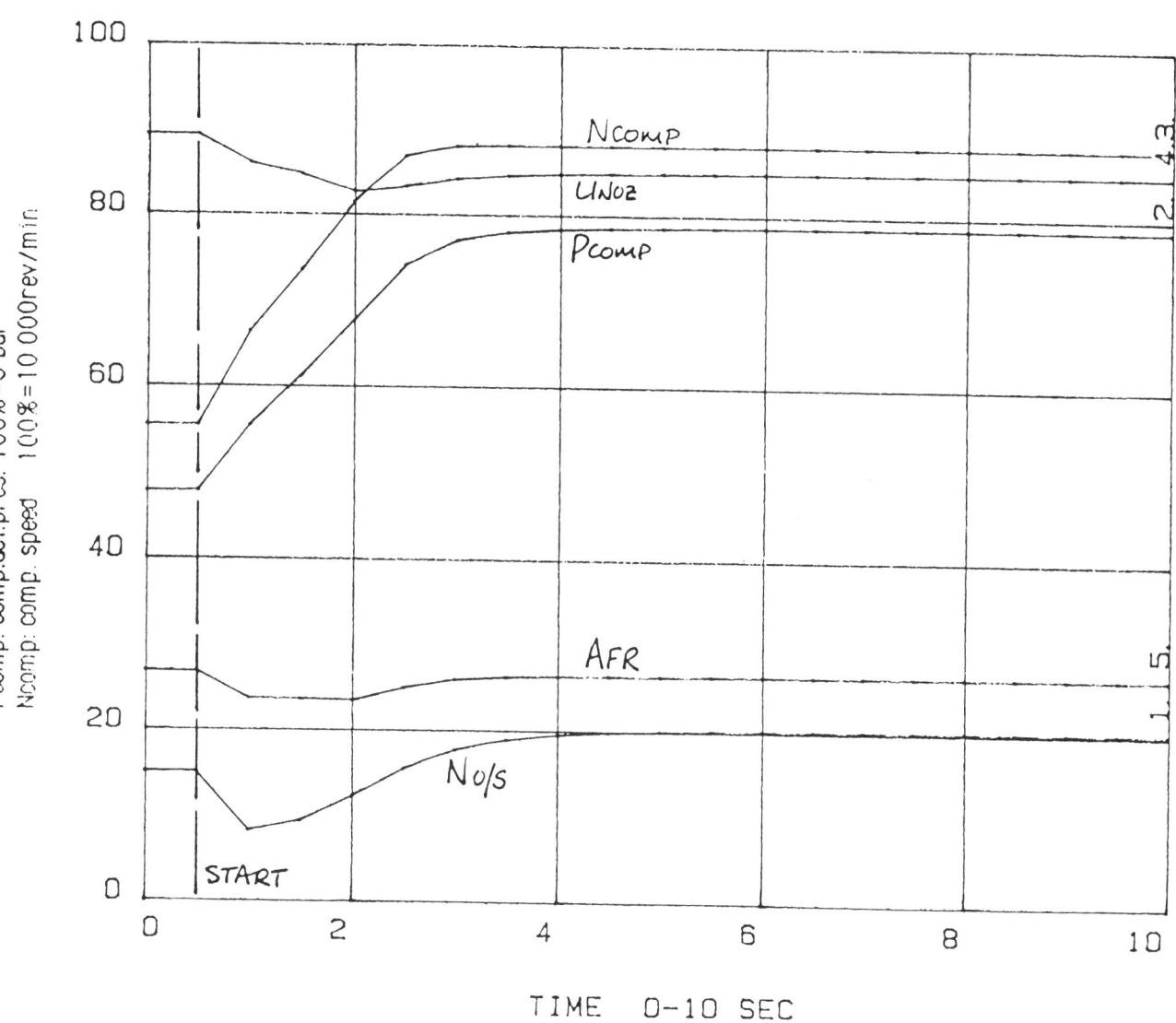

Fig 11c Combined step: C3

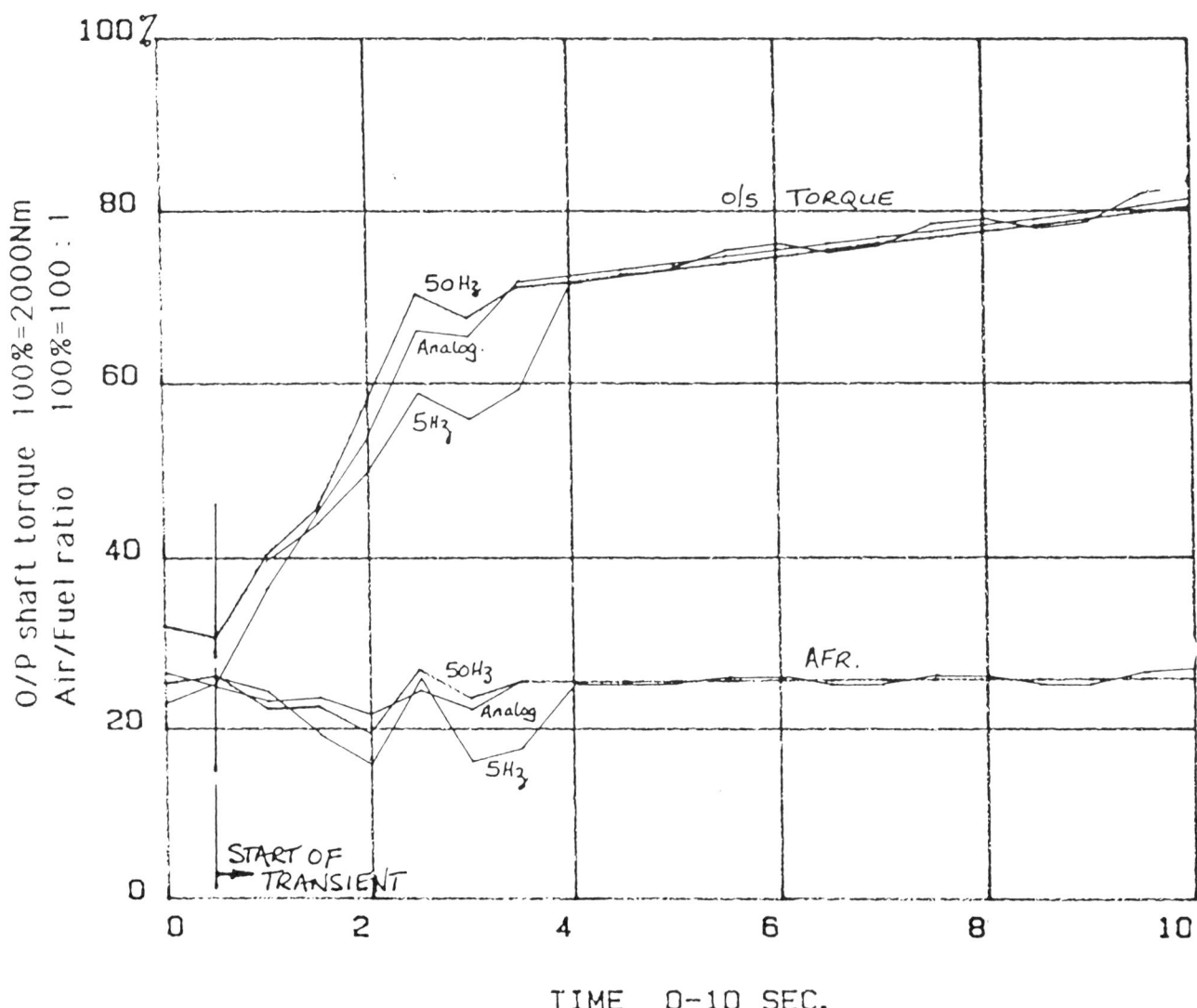

Fig 12 Load step: C2, discrete control

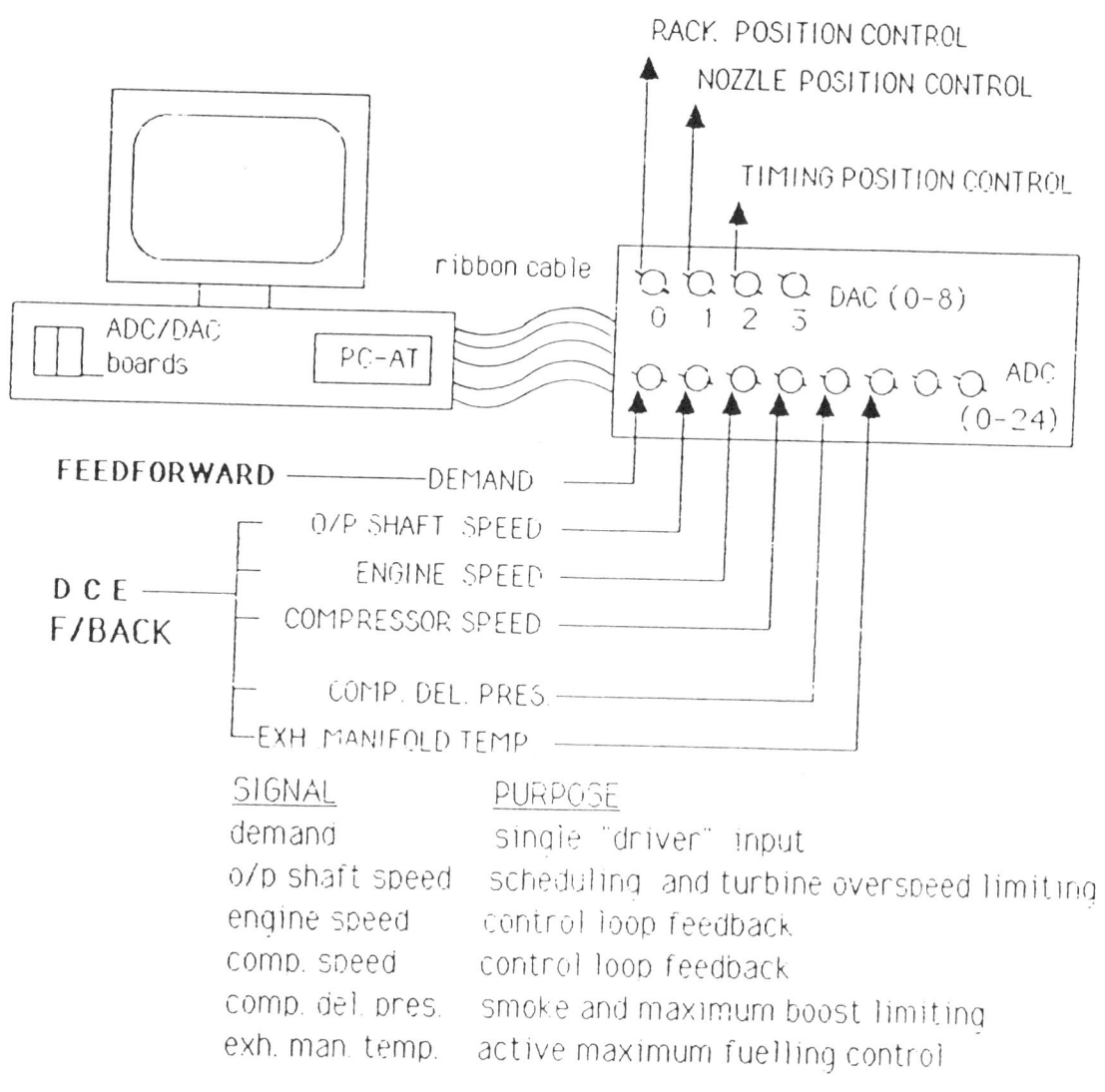

Fig 13   micro-controller hardware

890393

Transient Response and Route Simulations for Heavy Vehicles with Alternative Engine-Transmission System

Zou Dequan, M. Rezaian, and F.J. Wallace
School of Mechanical Engineering
University of Bath

ABSTRACT

The paper describes the results of a study of the dynamic response of a 30 tonne truck equipped with two alternative engine-transmission systems, viz.
a) the stepless differential compound engine (DCE)
b) a highly turbocharged engine of the same rating with a 7 ratio stepped transmission.

The comparisons cover :
a) 0 - 80 km/hr acceleration using a true dynamic and a quasi dynamic technique
b) a cross country 9 km route.

The DCE is shown to have superior acceleration characteristics but route behaviour of the two systems is similar.

INTRODUCTION

The Differential Compound Engine (DCE) has been proposed by Wallace et al as an alternative power unit for heavy vehicles, and performance and control aspects have been described in many publications (1-10), most recently in references 9 and 10. Its essential characteristic is that it acts, at one and at the same time, both as an integrated engine-transmission system and as a compounded Diesel engine with feedback of any excess power from the geared exhaust turbine over that required by the geared supercharging compressor. The layout of the scheme is shown in diagrammatic form in fig. 1, the essential features being :
a) the highly supercharged Diesel engine E
b) the fully floating epicyclic geartrain ECG in which the engine E drives the ring gear, while the sun gear and planet carrier are respectively connected to the supercharging compressor C and the output shaft.
c) the compressor (of the rotary positive displacement type)
d) the power turbine PT equipped with variable nozzles as an essential control device both for continuous steady state optimization and for best transient response.
e) the continuously variable turbine transmission CVT, to enable the turbine to operate at best efficiency under all operating conditions.
f) the controlled engine bypass BV to allow excess air to bypass the engine, but rejoin the engine exhaust gas stream upstream of the turbine.
g) the transmission torque converter TC with lock up facilities beyond 20% max. vehicle speed.
h) the control microprocessor MP to perform the continuous optimization as described in the companion paper (ref. 11).

Ref. 9 described the laboratory DCE at Bath University, based on the Leyland 520 6 cylinder DI Diesel engine having a swept volume of 8.2ℓ.

The present comparative study based on a 30 tonne (65000 lb) truck uses a larger 6 cylinder Diesel engine having a swept volume of 10ℓ and capable of sustained operation at high ratings; delivering 240 kW at 2100 rpm (BMEP = 13.69 bar). The epicyclic gear train, power turbine and compressor are similar to those described in ref. 9.

The alternative power plant considered is the same engine in turbocharged, aftercooled form, somewhat uprated from its standard operating conditions, to deliver approximately the same power, i.e. 240 kW, as in DCE form, again at 2100 rpm and driving the same 65000 lb. truck through a 7 speed gearbox.

The leading particulars of the two engine transmission systems are given in Tables 1 and 2.

The comparative analytical study of the two systems is based on :

a) acceleration on a level road from 0-80 km/hr. This, in turn, is subdivided into two sections -
  i) a true transient analysis involving the response of the internal engine/system components to the suddenly increased demand for power.
  ii) a quasi transient analysis in which the power unit moves instantaneously to the limiting torque curve, with interruptions for gear change in the case of the turbocharged engine.
b) a complete route simulation for a 97 km cross country route involving relatively frequent gear changes and fairly steep gradients.

The analytical procedures will be described in detail under the appropriate headings.

DETAILS OF ENGINE TRANSMISSION SYSTEMS
(Tables 1 and 2)

TABLE 1
DCE Specification
Engine :-
6 cylinder DI Diesel engine
swept vol. 10.0ℓ
rated speed 2100 rev/min
rated power 240 kW
max. torque speed 1350 rev/min.
max. torque power 240 kW

Compressor :-
Rotary positive displacement type
max. output 49 kg/min. (at max. torque)
rated output 26 kg/min. (at rated)
max. press. ratio 4.219:1
rated press. ratio 2.965:1
max. power absorption 181.8.8 kW
rated   "   "   63.0 kW
max. speed 9985 rev/min.
rated speed 5515 rev/min.

Turbine :-
Radial inflow with adjustable nozzles
rated output 74.96 kW
max. torque output 144.13 kW
rated pressure ratio 2.861:1
max. torque press.ratio 4.115:1

Epicyclic gearbox type :-
Fully floating

Ratios :-
engine to annulus 1:1
compressor to sunwheel 3.415:1
output shaft to planet carrier 1.982:1

Turbine CVT :-
Variable ratio between turbine and output shaft to maintain turbine at max. efficiency under all operating conditions

Output shaft conditions :-
Max. speed and power 1022 Nm at 2200 rev/min.
              (235.6 kW)

Max. torque   4046.9 Nm at 440 rev/min.
              (186.55 kW)

Transmission torque converter:-
Operative o/s speed range 0-440 rev/min.
Stall torque ratio 3.75 :1
Lock up   above 440 rev/min.

Back axle ratio   3.99:1

TABLE 2
Turbocharged Engine Transmission System Specification
Engine :-
6 cylinder DI Diesel engine
swept volume   10.0ℓ
rated speed   2100 rev/min.)substantially
rated power   249 kW   )uprated from
max. torque speed 1400 rev/min.)standard
max. torque power 246 kW   )conditions

Turbocharger :-
Single stage, radial inflow turbine, centrifugal compressor
rated speed   108,130 rev/min.
rated press. ratio   2.974:1
max. torque speed   101,236 rev/min.
max. torque press. ratio 2.838:1

Gearbox ratios :-
1st 10.13:1   5th 1.84 :1
2nd 5.99:1   6th 1.35:1
3rd 3.56:1   7th   1:1
4th 2.57:1

Back axle ratio :-
3.70:1

TABLE 3
Truck Specification
All up weight :-
65,000 lb = 29484 kg
vehicle width   2.256 m
vehicle height   3.505 m
wheel dia.   0.5011 m

Road Resistance Model
(smooth concrete or asphalt)

R=0.003633g (mass kg)+0.0324ρ (area m^2) (vel $\underline{km}$)^2

$+ \frac{\text{\%age gradient}}{100}$ g(mass kg)

TRUCK ACCELERATION 0-80 km/hour

FULLY TRANSIENT MODEL - As already stated, the fully transient model allows not only for vehicle response, but also for the response of the internal components of the engine-transmission system, e.g. epicyclic gear train acceleration for the DCE and turbocharger response for the 'conventionally' powered truck.

In the fully transient program, the system starts from an initial equilibrium state (vehicle at rest, engine idling) and accelerates gradually, following application of a fuel step.

The latter is modulated, both for the DCE and the turbocharged engine to prevent air-fuel ratio falling below 20:1. The program then proceeds cycle by cycle, by solving a set of dynamic equations for the complete mechanical system, including the vehicle itself, as well as the conservation equations for mass and energy for the engine cylinders and the inlet and exhaust manifold. The latter is treated as a constant pressure manifold for the DCE and a pulse manifold for the turbocharged engine. Solution of this system of equations results in updated values of powers, torques and speeds for each component - engine, compressor, turbine as well as pressures, temperatures and mass flow throughout the system.

In the case of the DCE acceleration is continuous while for the turbocharged engine, gear changes occur whenever the engine has reached its maximum speed in the current gear, as represented by fig. 8.

Fig. 2 shows the variation of transmitted power vs. time for the two systems. The ability of the DCE to transmit power continuously throughout the acceleration period, as against the intermittent power delivery of the turbocharged engine due to gear changes and the need to restrict engine fuelling during turbocharger acceleration in first and second gear are clearly shown. However, it is also noteworthy that peak power levels for the turbocharged engine in the higher gears consistently exceed those for the DCE, largely due to gear and negative compounding losses. Fig. 2 also indicates the compounding effect in the DCE i.e. $(P_{turb} - P_{comp})$, which in fact during the acceleration period is negative, and the gear losses running at over 20 kW throughout. However the torque converter power loss is limited to the very short period during which the latter is operative, i.e. only 2.2 sec.

Fig. 3 shows engine torque for the two systems and illustrates the serious deficiency of the turbocharged engine in first and second gear, as opposed to the continuous high torque of the DCE engine. However, later in the process the DCE shows a torque deficiency for the reasons already stated.

Fig. 4 shows the turbocharger speed variations on a base of vehicle speed which in turn account for the need for boost limited fuelling, especially in the lowest gears.

Finally, figs. 5a and 5b show respectively vehicle speed vs time for the full acceleration period and cumulative fuel consumption vs time up to a vehicle speed of 30 km/hr only.

It is clear from fig. 5a that the DCE powered truck has a considerable advantage, reaching 80.6 km/hr after 54.0 sec. as against 61.3 sec. for the turbocharged truck.

Fig. 5b shows that w.r.t cumulative fuel consumption up to 30 km/hr, the DCE again has a significant advantage, with a figure of approx. .07 kg as against .08 kg for the turbocharged truck. However, in the later stages of acceleration, the gear losses of the DCE exert an adverse effect, so that for the 0-80.6 km/hr acceleration cumulative consumption for the DCE is 0.675 kg for the DCE as against 0.652 kg for the turbocharged truck.

QUASI-TRANSIENT MODEL - In this model, both power plants are assumed to respond instantaneously to increased demand of torque. Both the DCE and the turbocharged engine thus reach steady state limiting torque corresponding to the initial output shaft or engine speed without delay. In both cases acceleration of the complete vehicle then proceeds under the action of limiting torque until the final vehicle speed is reached. Subsequently fuelling and hence output torque are reduced to the level required to maintain this final vehicle speed against the road resistance. This model is also important because it is the basis of the route simulation described in Section 3.3.

The steady state characteristics of the fully optimized DCE are shown in figs. 6a,b and c; giving contours of
a) overall i.e. system efficiency
b) boost pressure ratio, and
c) engine BMEP
with output shaft torque as the ordinate and output shaft speed as the abscissa (see also table 1).

Fig. 6a shows the relatively high system efficiencies which have to be compared with the combined efficiency of the turbocharged engine and gearbox. The continuous torque rise with decreasing output shaft speed is equally apparent. As indicated in table 1, a normally locked up transmission torque converter is used for the lowest fifth of the vehicle speed range, increasing the overall torque ratio between stall and rated speed [2200 rev/min.= 106.5 km/hr.] to approx. 14.8:1.

Fig. 6b shows contours of boost pressure ratio and fig. 6c shows contours of engine BMEP following a very similar trend to those of boost, both rising steadily on the limiting torque curve towards the low speed end.

Fig. 7 shows the equivalent tractive effort vs road speed relationship for the DCE on the limiting torque curve superimposed on the equivalent tractive effort curves of the turbocharged engine in the various gears.

The calculated acceleration times and fuel consumption up to a speed of 80.6 km/hr for the two systems, calculated on the basis of quasi-transient behaviour are now as follows :-

TABLE 4

	DCE truck	T/C truck
Acceleration time (secs)	52.4 (54.0)	56.5 (61.32)
Cum. fuel consumption kg	.655 (0.675)	0.690 (0.652)

(Figures in parenthesis represent corresponding results for 'true transient' acceleration).

Table 4 shows the clear advantage of the DCE over the turbocharged engine both for the true transient and quasi-transient case, in terms of acceleration time. It also shows that, whereas for the DCE the differences between the two methods are comparatively slight, with only marginal reduction in acceleration time and fuel consumption for the quasi-transient case, the differences are much greater in the case of the turbocharged engine where engine torque deficiency in the lower gears due to turbocharger lag, makes the true acceleration time nearly 5 secs. longer than the quasi-transient period, as against a difference of only 1.6 secs. for the DCE. With regard to cumulative fuel consumption, similar remarks apply except for the fact that, for the DCE, the situation changes from an adverse one for the true transient [0.675 kg for the DCE cf. 0.652 kg for the T/C engine] to a favourable one [0.655 kg for the DCE cf. 0.690 kg for the T/C engine] for the quasi transient.

These comparisons are also significant with respect to the route simulation in the next section which is based on the quasi-transient model and which will therefore tend to lead to optimistic predictions for total time, though not necessarily for cumulative fuel consumption.

ROUTE SIMULATION - For this purpose a typical UK route of 97 km length, with variable gradients up to 5% approx. was chosen. The simulation was carried out in terms of driver demanded speed as follows :

When demanded speed exceeds actual speed, the engine system immediately operates on the limiting torque curve as shown in fig. 7. In the case of the turbocharged engine the gear chosen is determined by the instantaneous vehicle speed, as shown in fig. 7, with engine speed expressed as a function of vehicle speed, as shown in fig. 8. In the case of the DCE, the available torque is again derived from fig. 7, with instantaneous fuel consumption in kg/kWhr. derived from the DCE operating map, fig. 6a, and similarly for the turbocharged engine. The program proceeds in fixed calculation time steps of 1 sec. until the required speed is reached, with gear changes at intermediate speeds for the turbocharged engine, as and when required. Instantaneous acceleration of the vehicle is obtained by taking the difference between available tractive effort, from fig. 7 and tractive resistance as given by the expression in Section 2.4, with due allowance for the instantaneous gradient. The distance travelled and fuel consumed are then calculated for each time step and progressively accumulated. When the desired speed is reached, the demanded tractive effort is reduced instantaneously to the appropriate tractive resistance.

In the case of downhill operation or reduced demanded speed, braking effort is applied, with the engine simultaneously moving to idle operation. In the case of the turbocharged truck, downward gear changes are performed in line with the schedule of fig. 8.

The flow chart for the complete route simulation program, in the case of the turbocharged truck, is shown in fig. A.1.

The route profile, in terms of gradients and desired speeds against distance, is shown in fig. 9; while the results, expressed as distance travelled and cumulative fuel consumption against time are shown in figs. 10a and 10b for the DCE and turbocharged truck, respectively. Fig. 10b also indicates the various gear changes, 30 in all.

The global comparisons between the DCE and turbocharged engine powered truck are summarized in table 5.

TABLE 5

	DCE truck	T/C truck
Route length (km)	97.0	97.0
total time (mins)	89.0	89.1
total fuel consumption (kg)	21.68	23.00
No. of gear changes	NIL	30

It is somewhat surprising that the total journey time is virtually identical in the two cases, although as expected there is a significant saving in fuel consumption for the DCE. Part of the explanation may be found in fig. 7 showing a definite tractive effort deficiency based on the assumption of similar engine ratings (see tables 1 and 2) and arising from negative compounding and high gear losses at low speeds compared with the conventionally powered truck. This will lead to a reduction of speed on long gradients, which on this route outweigh the gains in acceleration times shown in table 4. Nevertheless, bearing in mind the optimistic nature of the quasi transient acceleration predictions for the turbocharged truck, as also shown in table 4, one would expect to see some increase in real journey time for the latter, as compared with the DCE powered truck.

CONCLUSIONS

a) the DCE powered truck achieves significantly better acceleration times up to 80 km/hr than the equivalent turbocharged engine powered truck.

b) the normal method of simulation used for vehicle response, referred to here as the quasi transient method considerably underestimates the acceleration time for the turbocharged truck, by approx. 8%, due to neglect of lack of engine torque in the lower gears.

c) as a result of this latter neglect, route times for the turbocharged truck will also tend to be assessed optimistically, as against the DCE in which the transient lag is virtually absent.

d) the performance of the DCE is critically dependent on compressor and turbine efficiencies, as well as gear losses. Any improvements here will be reflected in substantially improved journey times and reduced fuel consumption.

e) it may be concluded that the advantages of the DCE system which may be described as marginal in the truck application, become more pronounced in military vehicle applications where, due to the inherently more favourable power balance conditions for compound engines at higher ratings, and the much more severe nature of the terrain, both fuel consumption and mission time will show sizeable gains over a vehicle powered by a very highly turbocharged engine.

ACKNOWLEDGEMENTS

The authors are indebted to the Cummins Engine Company for financial support and permission to publish this paper.

REFERENCES

(1) WALLACE, F.J.
"Operating Characteristics of Compound Engine Schemes for Traction Purposes based on Opposed Piston Two Stroke Engines with Differential Gearing",
Proc.IMechE, Vol.177, No.2, 1963.

(2) WALLACE, F.J.
"Theoretical and Experimental Analysis of Air and Gas Flows in a Crankcase Scavenged Two stroke Engine Employing Boost Ports", SAE Detroit, 1969, Paper 690134.

(3) WALLACE, F.J., WINKLER, G. and BOWNS, D.E.
"Multi-Variable Control of Diesel Engine/ Transmission Systems with Infinitely Variable Ratios".
SAE Congress, Detroit, Feb. 1977, Paper 770125.

(4) WALLACE, F.J. and WINKLER, G.
"Very High Output Diesel Engines - A Critical Comparison of Two Stage Turbocharged, Hyperbar and Differential Compound Engines",
SAE, Milwaukee, No. 770756, Sept. 1977.

(5) WALLACE, F.J. and KIMBER, R.M.
"Optimization of the Differential Compound Engine using Microprocessor Control",
SAE, Detroit, Paper 810336, Feb. 1981.

(6) WALLACE, F.J., TARABAD, M. and HOWARD, D.
"Steady State and Transient Control through a Microprocessor-Electrohydraulic System of an Integrated Engine-Transmission Unit",
SAE, Detroit, Paper 830578, March 1983.

(7) WALLACE, F.J., TARABAD, M. and HOWARD, D.
"The Differential Compound Engine - a New Integrated Engine Transmission System Concept for Heavy Vehicles",
Proc.IMechE, Vol. 197A, 47/83.

(8) WALLACE, F.J. and TARABAD, M.
"Engine Transmission Systems for the Heavy Goods Vehicle and the Passenger Carrying Bus", Proc.IMechE, Vol. 197A, 31/83.

(9) WALLACE, F.J., PRINCE, D., HOWARD, D. and TARABAD, M.
"Design and Performance Characteristics of the Laboratory Differential Compound Engine

(10) WALLACE, F.J., TARABAD, M. and HOWARD, D.
"Design and Performance Studies for a 1000 h.p Military Version of the Differential Compound Engine", IMechE C194/86.

(11) HALL, J. and WALLACE, F.J.
"Control Design for a Differential Compound Engine", to be presented at the SAE Congress Detroit, February 1989.

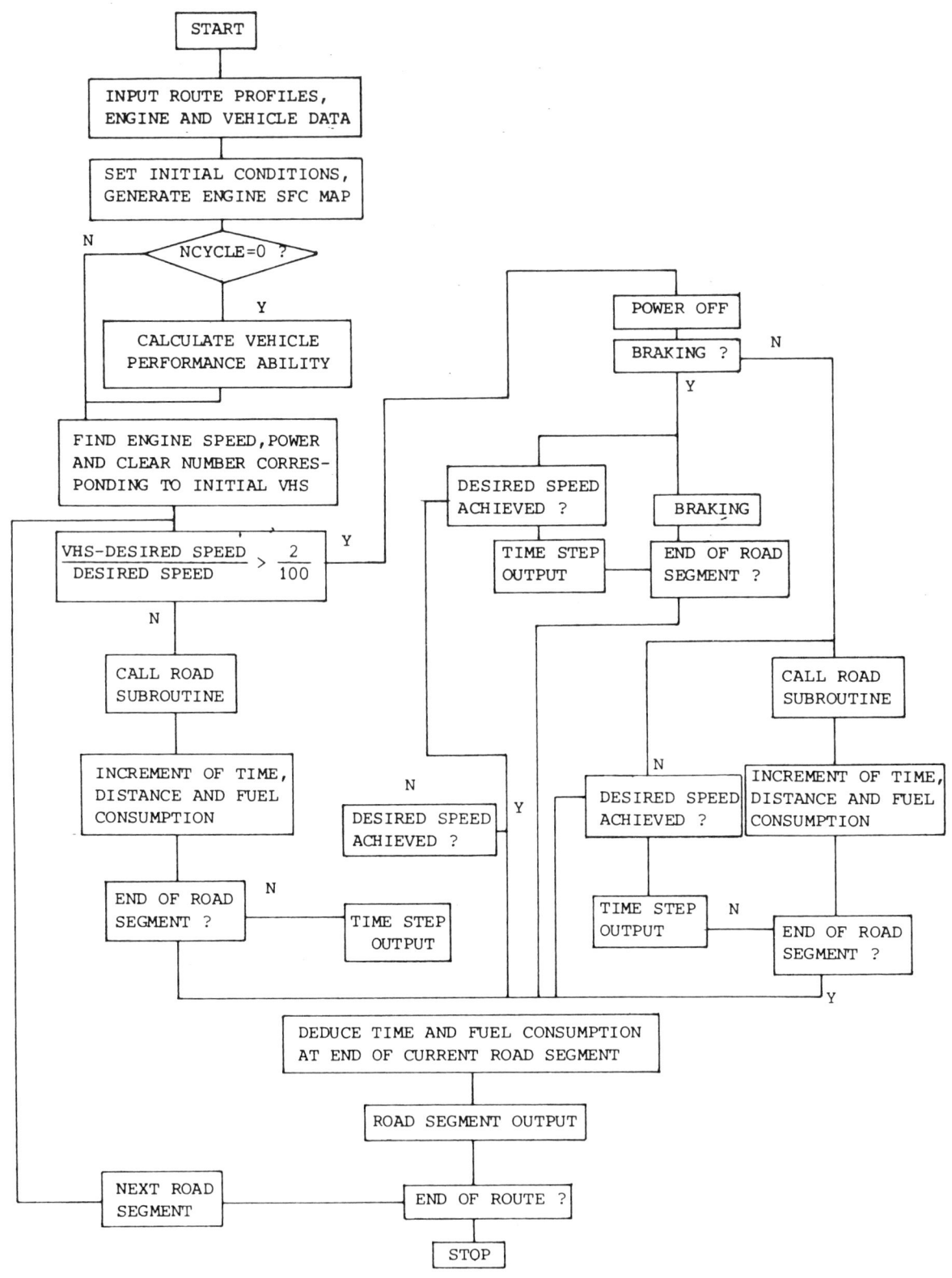

FIG. A1

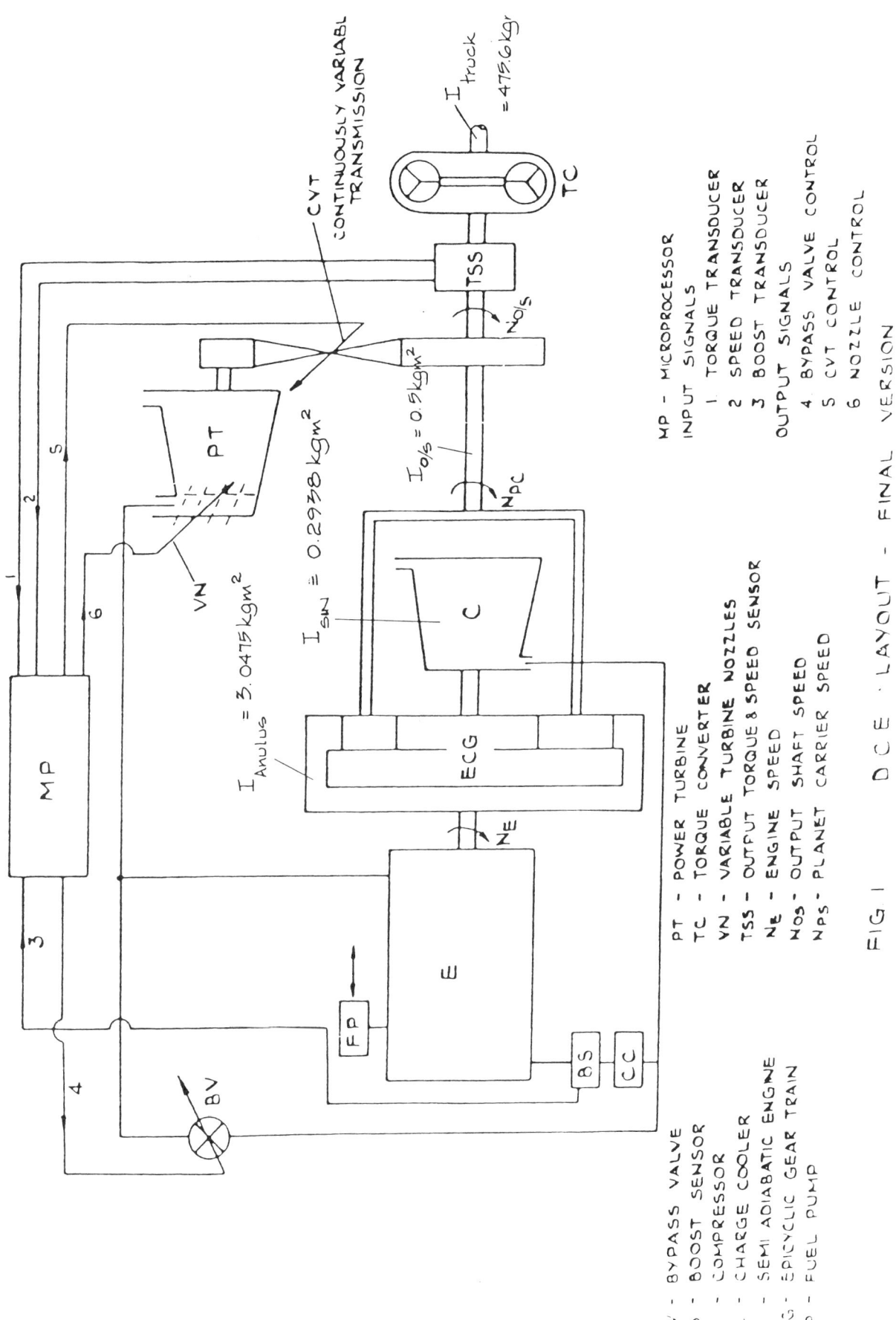

FIG 1 D C E - LAYOUT - FINAL VERSION

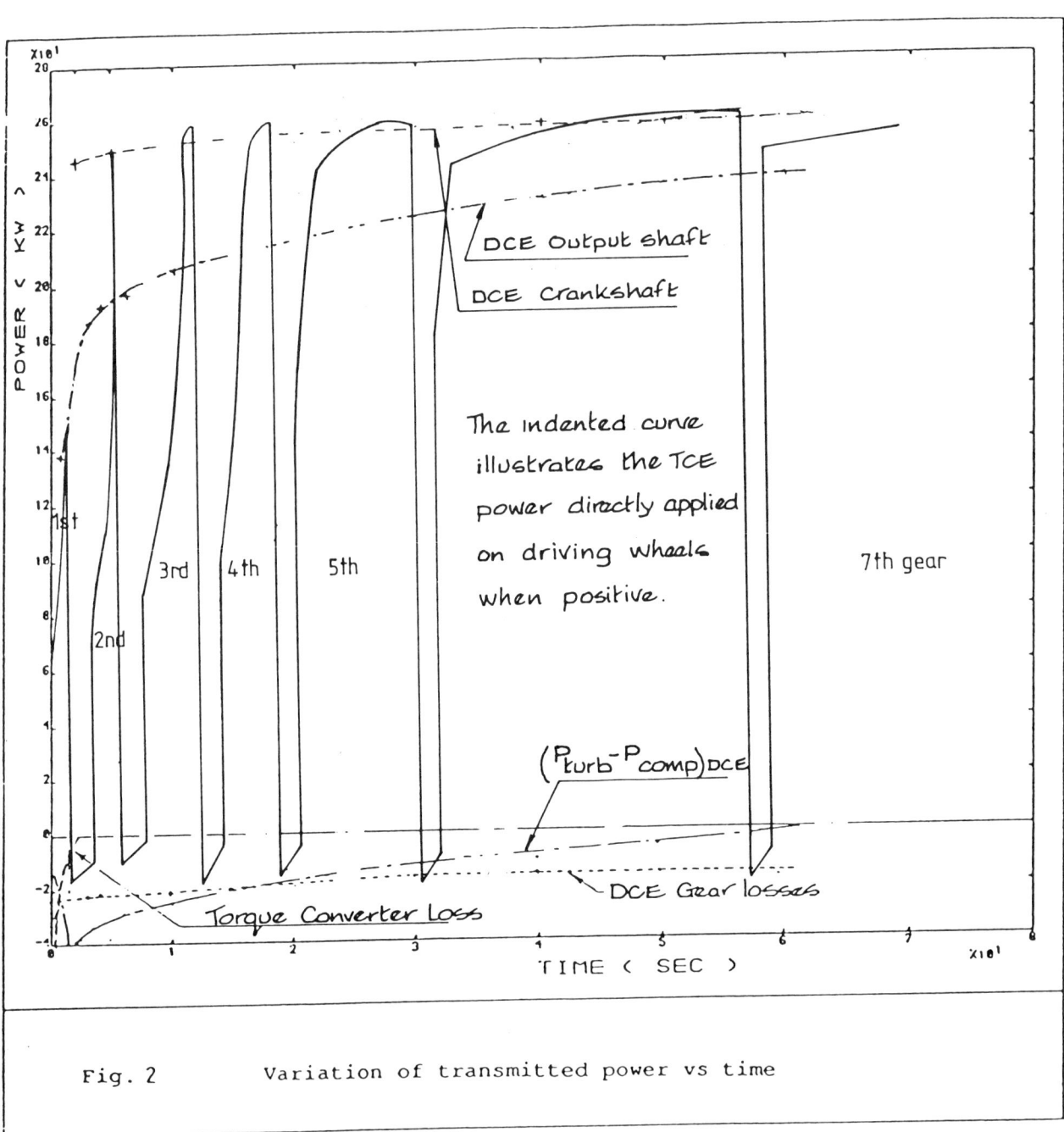

Fig. 2   Variation of transmitted power vs time

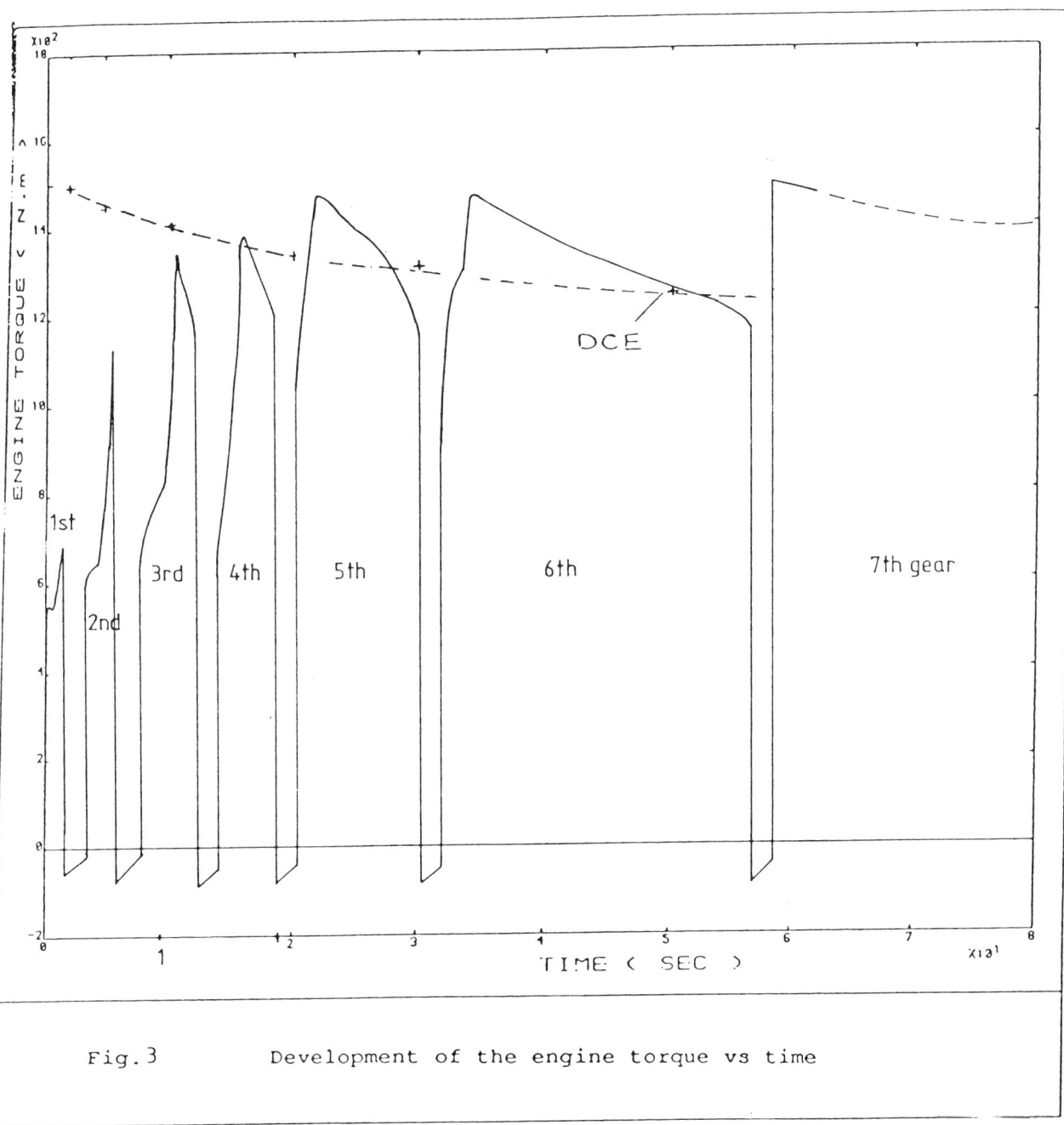

Fig. 3   Development of the engine torque vs time

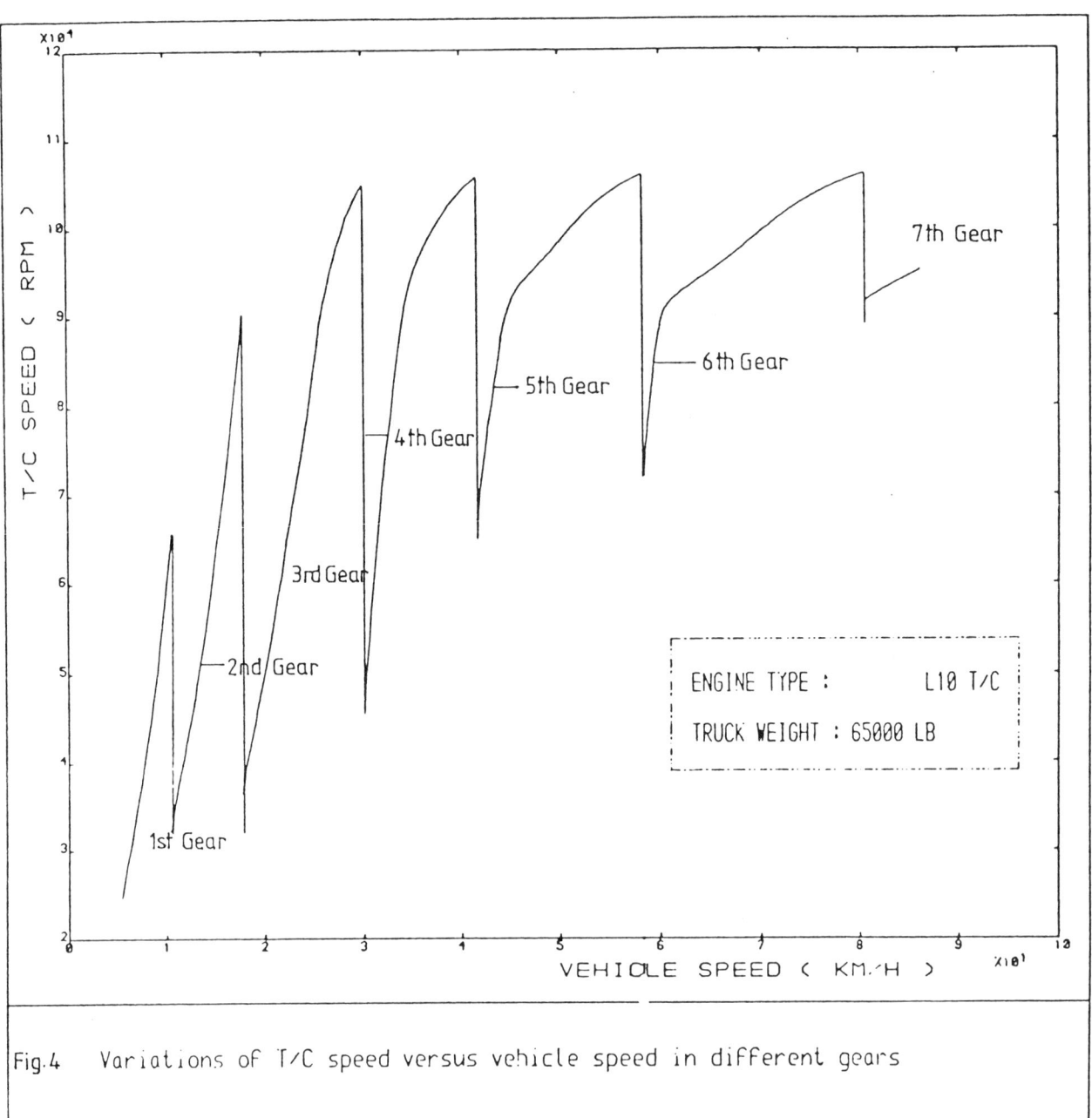

Fig.4 Variations of T/C speed versus vehicle speed in different gears

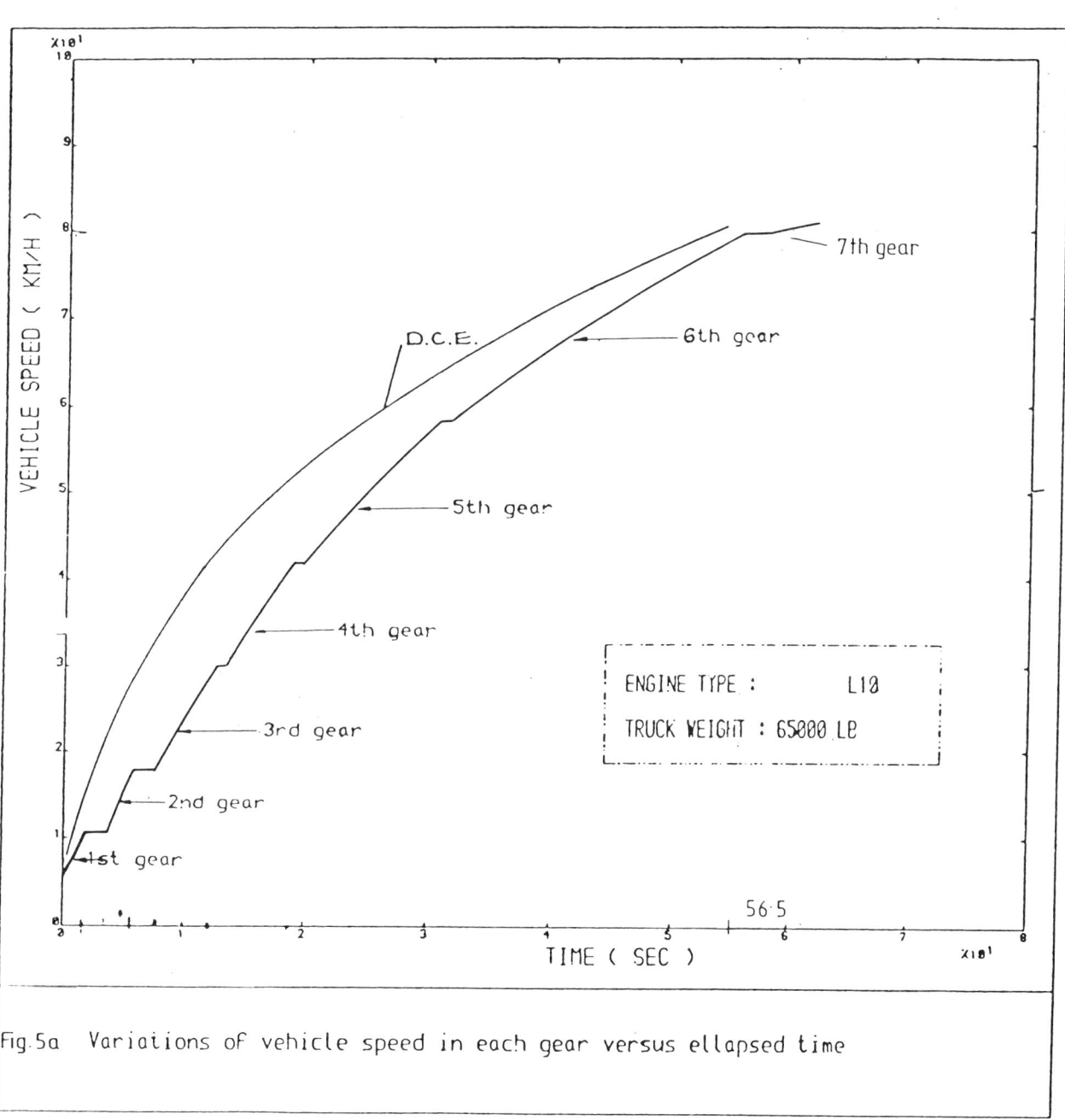

Fig.5a Variations of vehicle speed in each gear versus ellapsed time

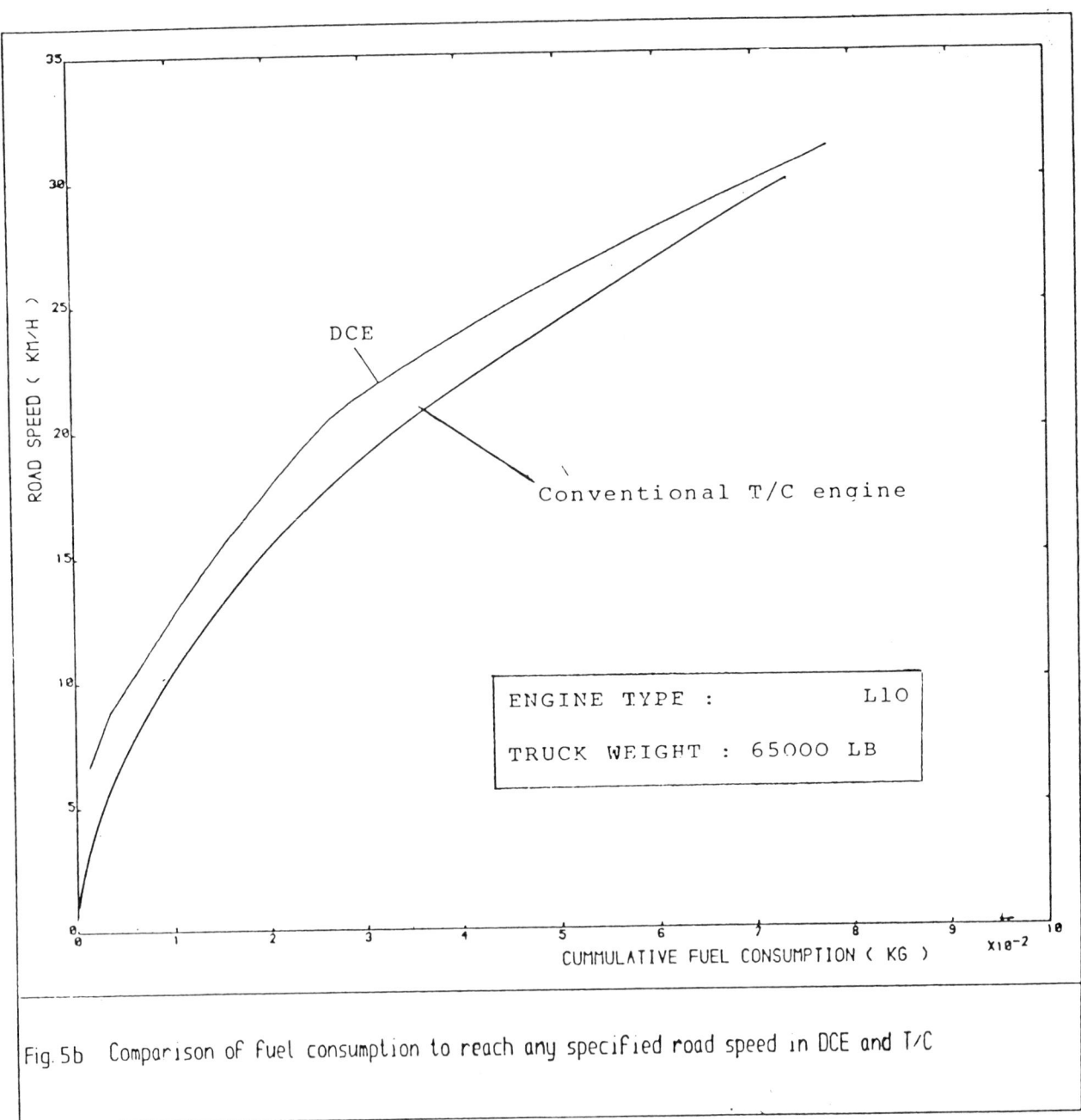

Fig. 5b Comparison of fuel consumption to reach any specified road speed in DCE and T/C

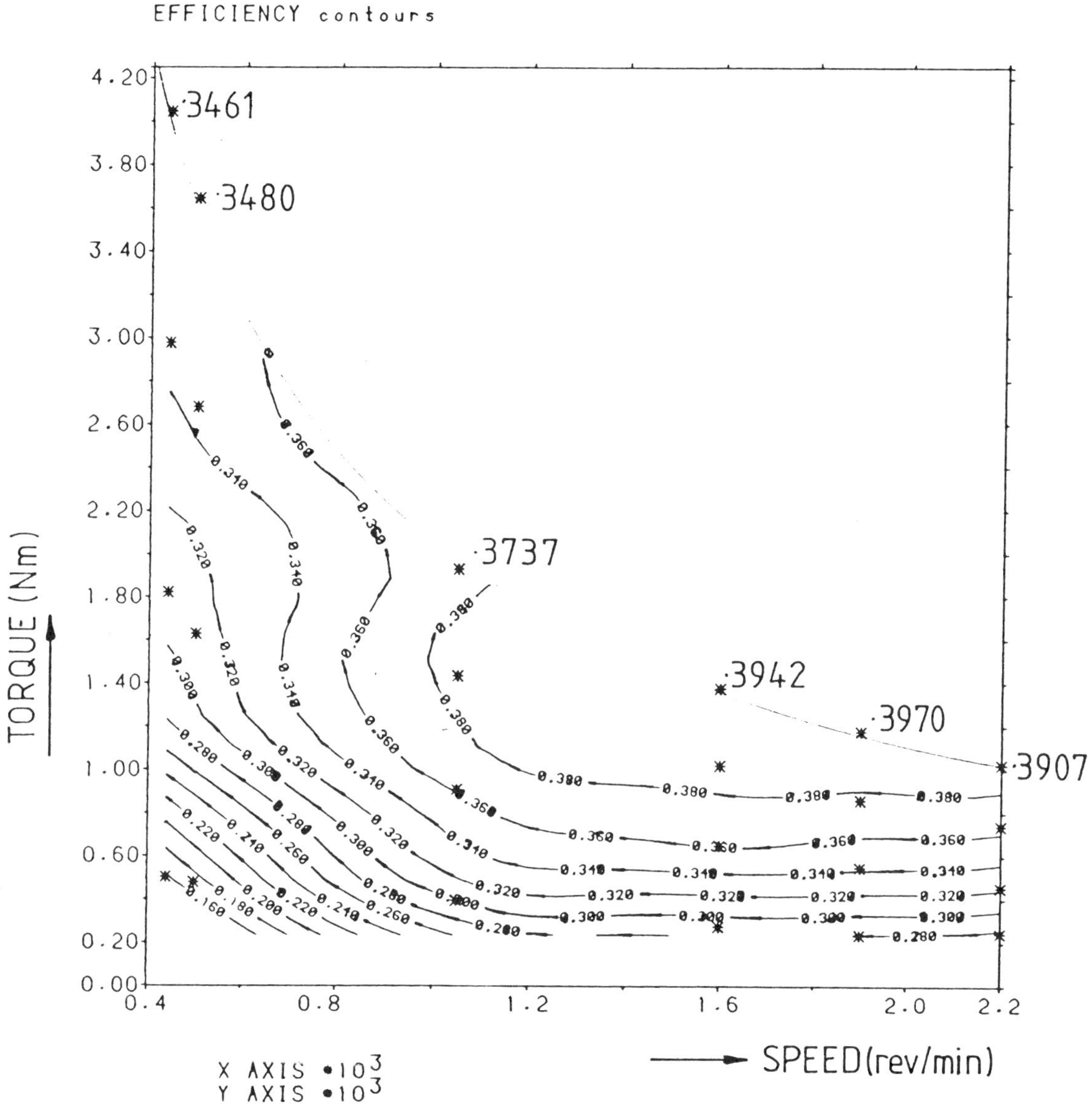

Fig.6a EFFICIENCY CONTOURS
CR = 16.3

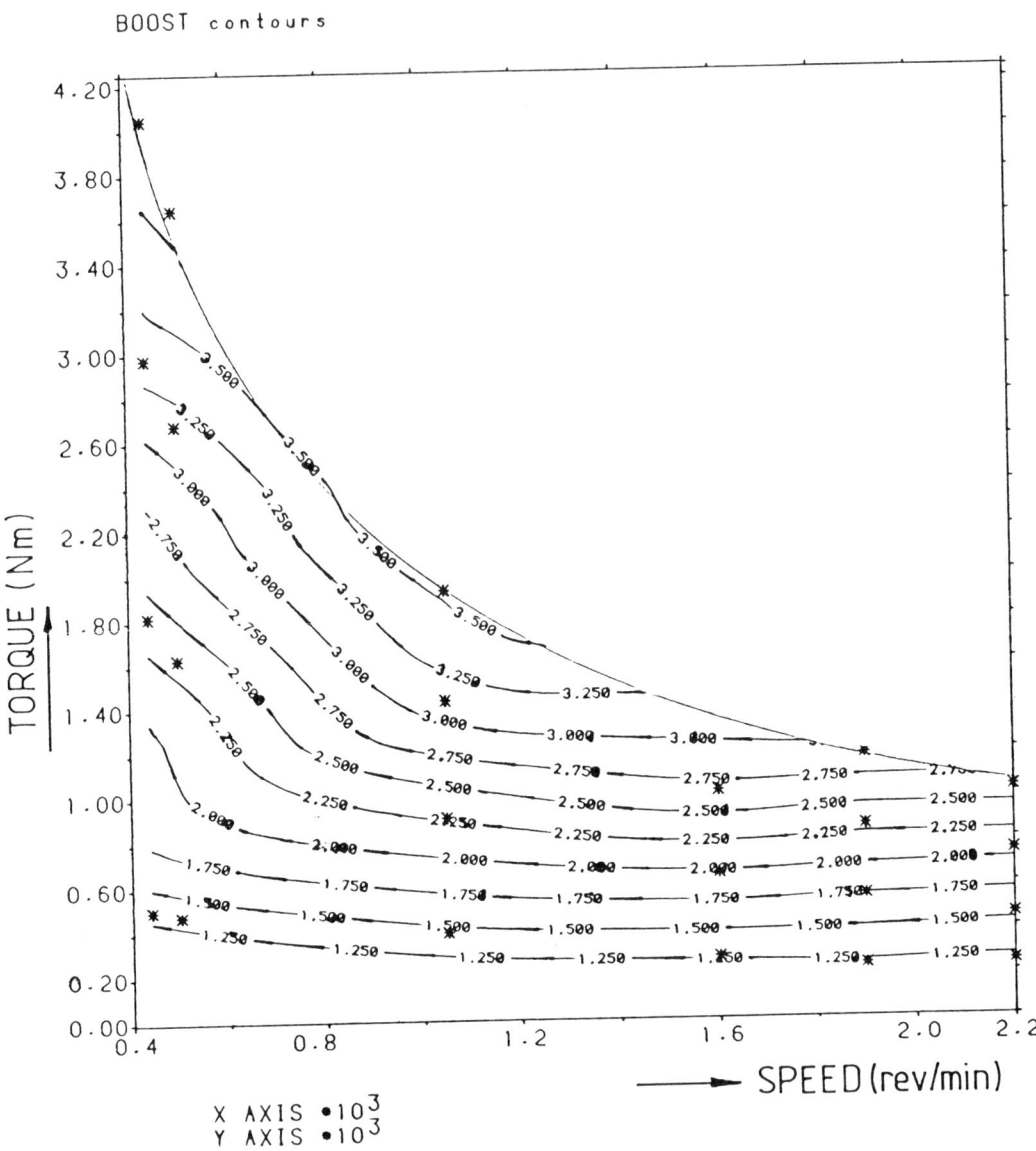

Fig. 6b BOOST CONTOURS CR = 16.3

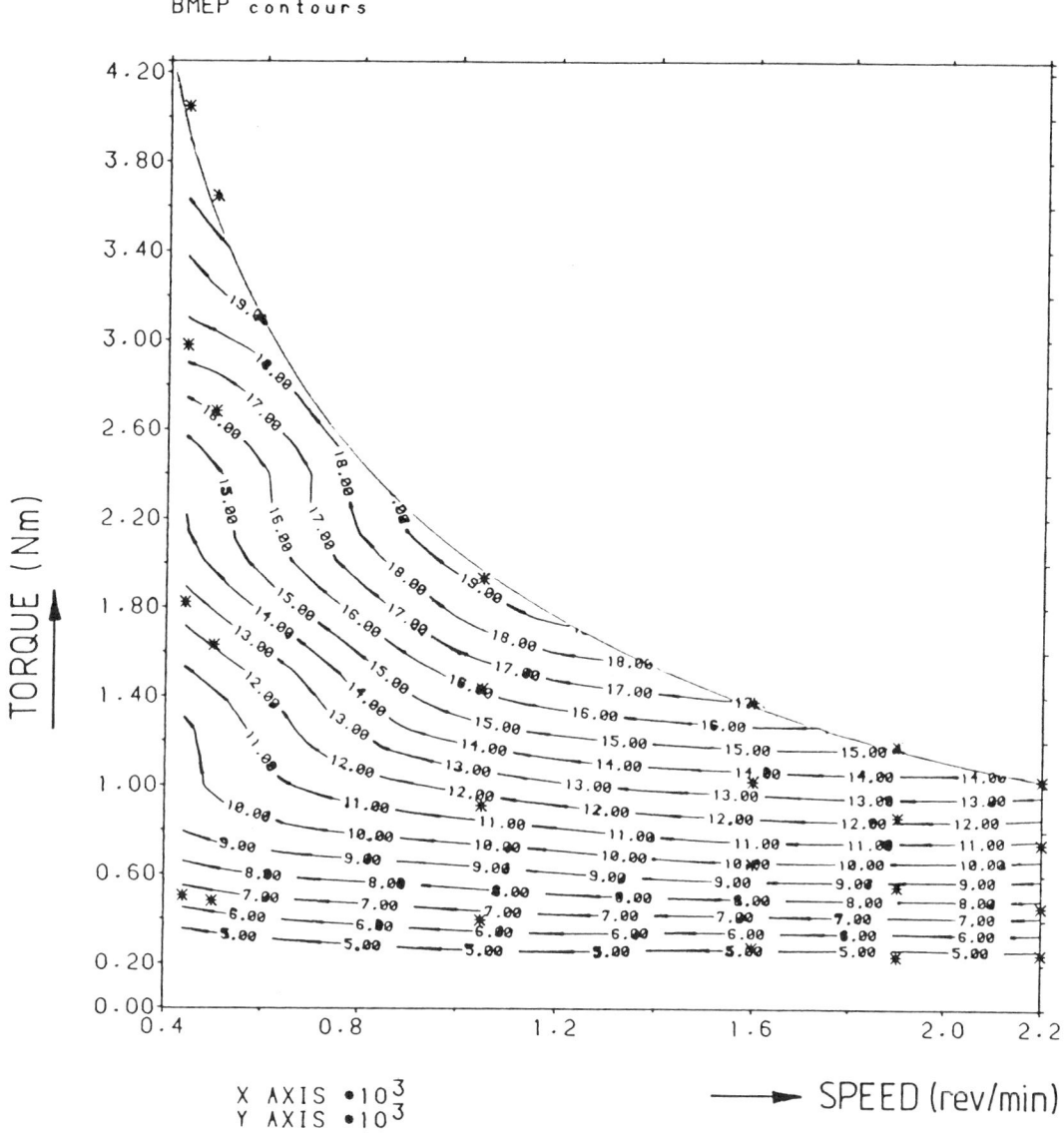

Fig. 6c BMEP CONTOURS
CR = 16·3

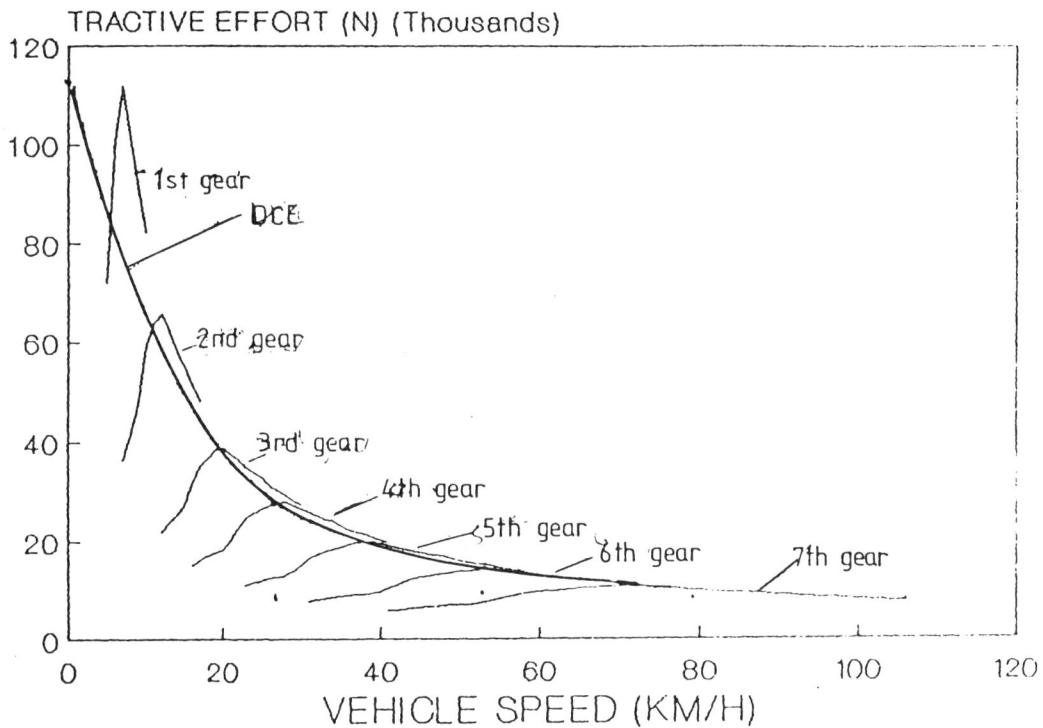

Fig.7 Tractive effort vs. speed

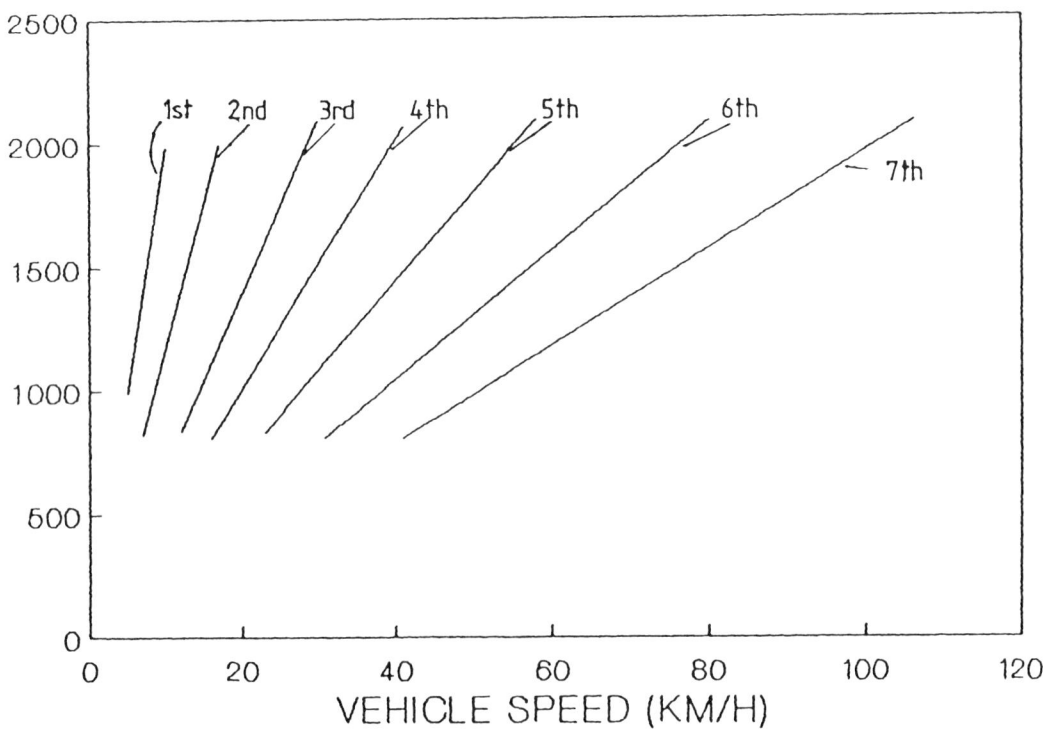

Fig.8   Engine speed vs. vehicle speed in different gears

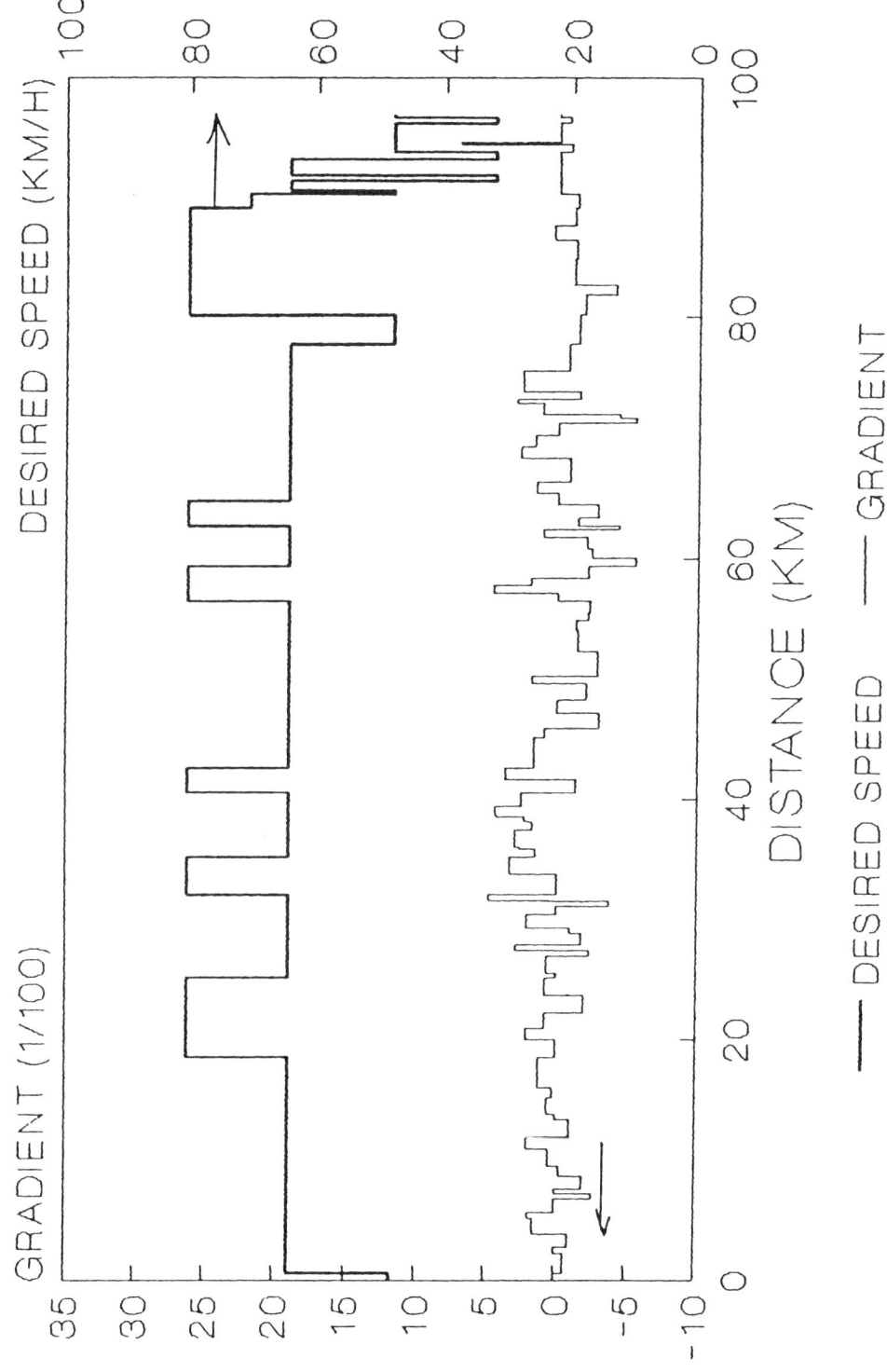

FIG.9 ROUTE PROFILE

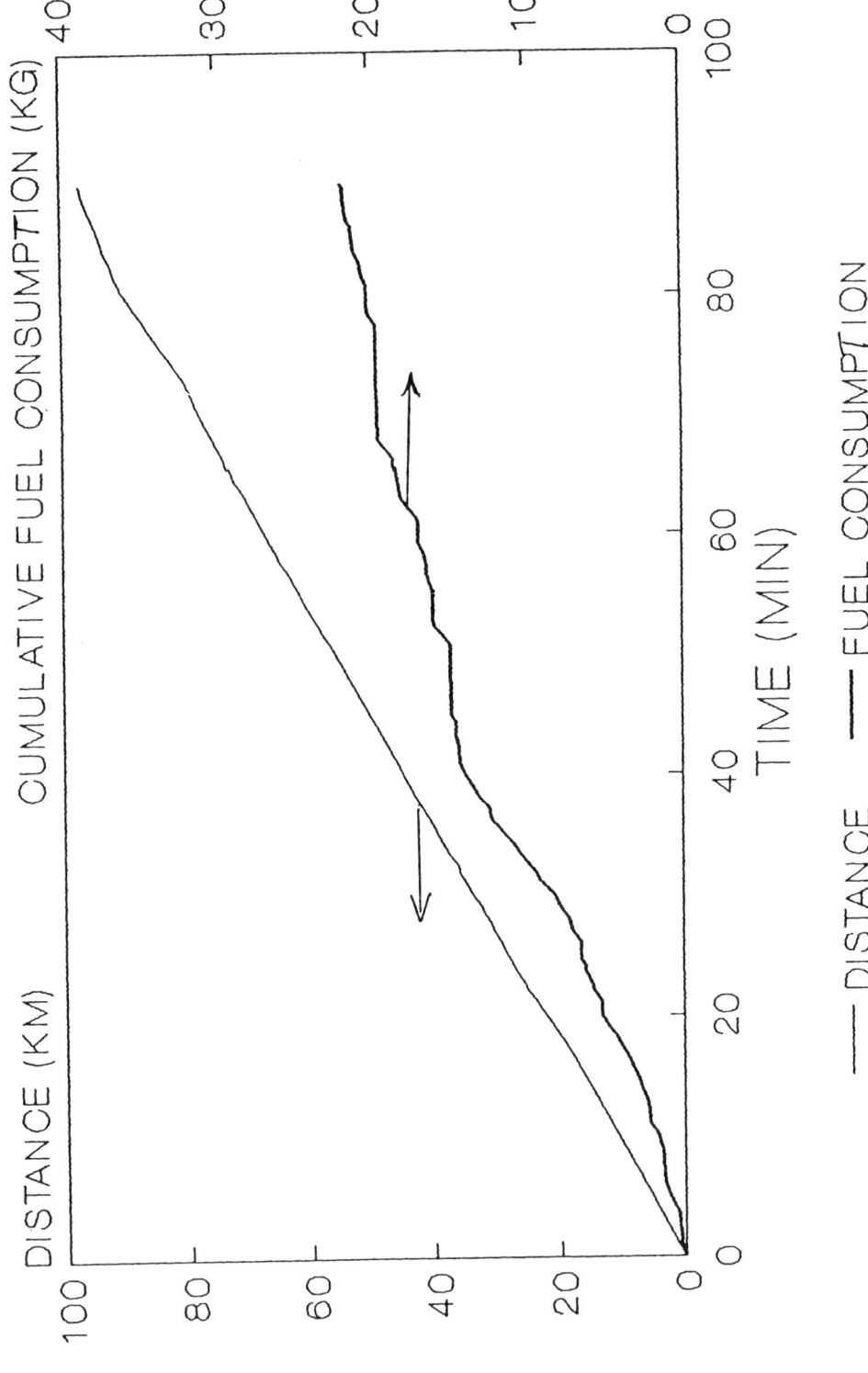

FIG.10a ROUTE PERFORMANCE--DOE TRUCK

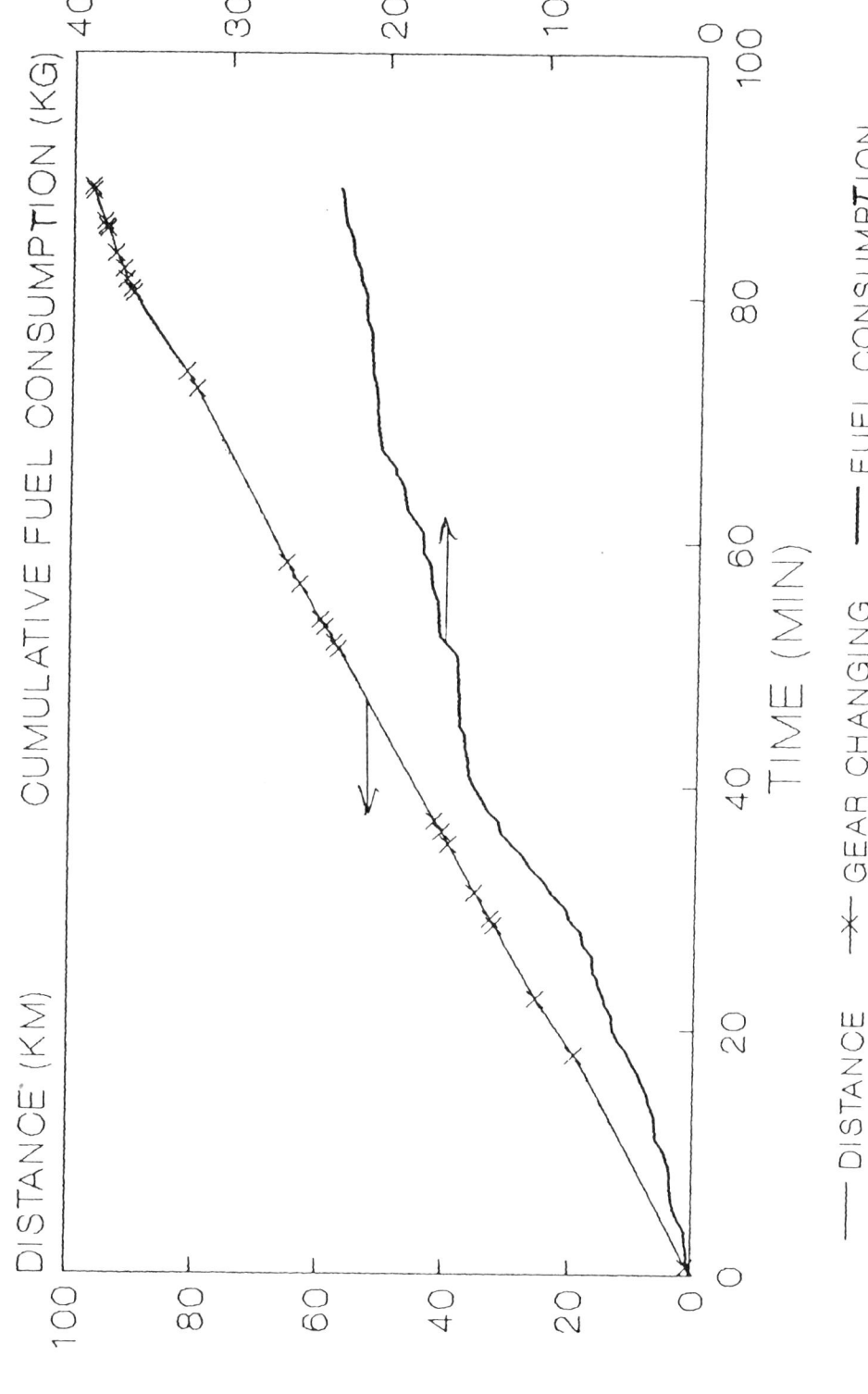

FIG.10b ROUTE PERFORMANCE--T/C TRUCK

890394

Dynamic Simulation of a Turbocharged/Intercooled Diesel Engine with Rack-Actuated Electronic Fuel Control System

Charles I. Rackmil and Paul N. Blumberg
Integral Technologies Inc.
Westmont, Il.

ABSTRACT

A comprehensive simulation has been developed which describes the dynamic response of a turbocharged, six cylinder diesel engine fueled by an in-line pump-line-nozzle system with electronic control of a solenoid-actuated rack. The mathematical models of components, governing strategies and control algorithms are general, having been developed to address issues related to the dynamic response and control of both components and the overall system in a variety of engine propulsion and power system applications. The physical principles and salient characteristics of the independent subsystem models are reviewed. Results of the overall system simulation, featuring interactions between the engine, microprocessor-based governor, and solenoid-actuated rack are presented to demonstrate the utility of dynamic simulation as a basis for evaluating system performance and exploring control strategies for optimum response.

RECENT PUBLICATIONS have demonstrated the successful development and application of comprehensive, physically-based, state space numerical simulations of engine-driven propulsion and power systems [1, 2, 3, 4][*]. In general, these simulations include detailed submodels of the engine, driveline or power transmission system, vehicle or load system, fuel controller/governor, sensors, actuators, microprocessors and other relevant control elements. Among the objectives typically associated with the development of these simulations are:

1) dynamic evaluation of the performance of the overall propulsion system under a variety of realistic operating conditions, reducing the required test burden and development time;

2) identification of key propulsion system components which influence system response and quantitative characterization of their effect;
3) exploration of alternative control structures and strategies, including determination of calibration parameters, for optimizing system performance under specific operating conditions;
4) specification, design and integration of sensors and actuators and other subsystem componentry.

We refer to time-based simulations of this type as dynamic simulations and to the computer-aided engineering methodology embodied in this approach as Dynamic Simulation Methodology (DSM).

The present work focuses on the design and control issues surrounding the central actuator in a diesel power or propulsion system, i.e., the fuel injection system, and its interaction with other aspects of engine control. In particular, we employ DSM to examine the performance of a diesel engine fueled by an in-line high pressure pump with a solenoid-actuated, electronically controlled rack. Using DSM, issues related to the design and control of the rack actuator, and its interrelationship with engine speed governing in different applications are analyzed and discussed.

PROPULSION/POWER SYSTEM DESCRIPTION

The structure of the present simulation is shown in Figure 1, which indicates the major components included in the simulation and their interactions. The engine considered is a four-stroke, turbocharged and intercooled diesel which accelerates or decelerates in response to varying output and/or load torques. The precise nature of the load torque is application dependent (e.g., vehicle driveline or generator set), but for the purposes of the current study is adequately described as a rigid load geared to the engine through a gear set of variable

---
[*]Numbers in brackets designate references listed at the end of the paper.

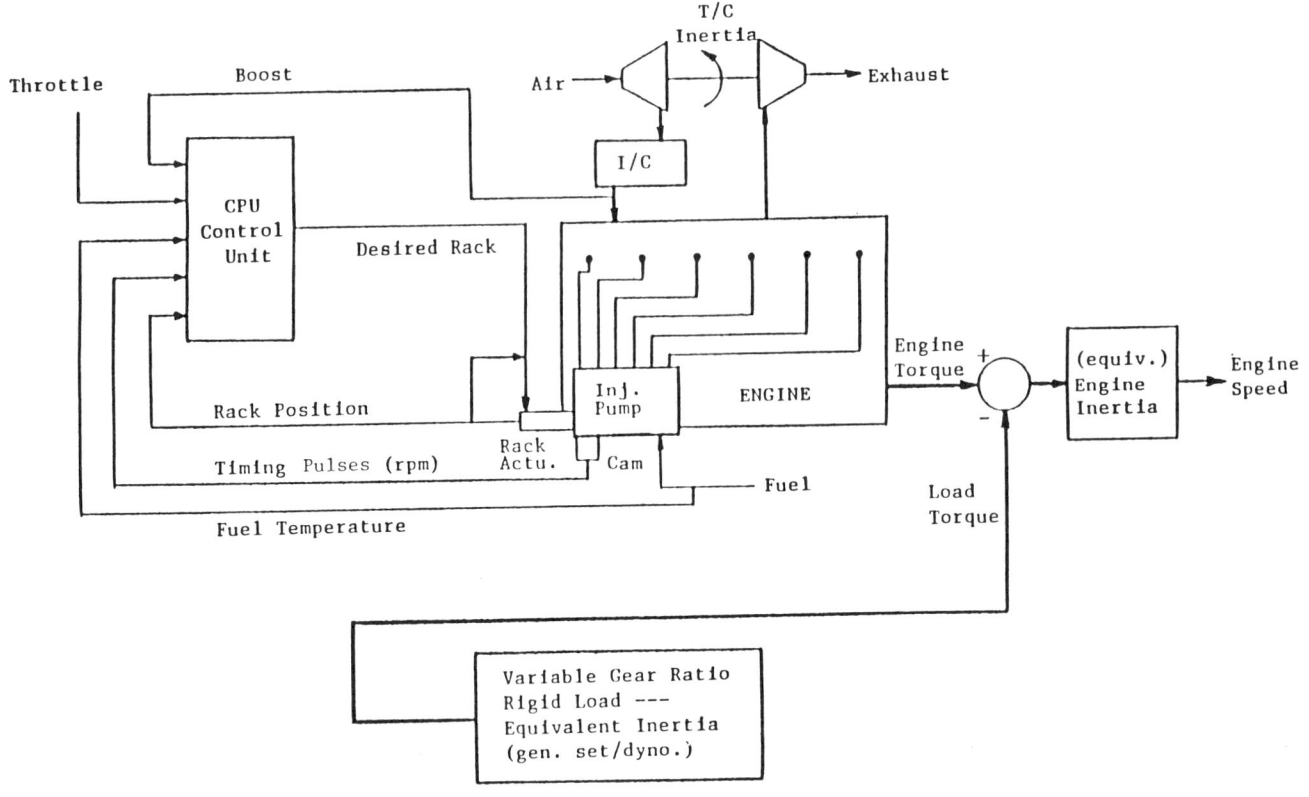

Figure 1. Schematic of a generalized propulsion/power system.

ratio. Crankshaft position is obtained from a sensor and an engine-driven toothed timing cam. The measured engine speed is then determined from the rate of change of crankshaft position. Desired fueling is determined through the microprocessor-based fuel controller and governor, which dynamically evaluates throttle position, desired versus measured engine speed, boost pressure and fueling limits (e.g., limits due to boost pressure and maximum torque). Desired rack position is determined from desired fueling and engine speed, as a function of the fueling characteristics of the pump-line-nozzle system. The rack is controlled to its desired value through a linear magnetic solenoid actuator with return spring, utilizing a rack position sensor for feedback, similar to that reported in the literature by several fuel injection equipment manufacturers [5, 6]. In the system analyzed, the rack control logic is assumed to be executed continuously through analog circuitry with no measurable delay, whereas the governor control is microprocessor-based and asynchronous with engine events.

The rack actuator and controller must provide responsive control of rack such that it closely approaches its time-varying desired value without unacceptable delays or overshoot. Rack performance should be relatively independent of operating conditions, primarily rack position, and, in addition, insensitive to any changes in the physical or mechanical properties of the rack subsystem due to such things as friction, spring "set", drift in solenoid force vs. current, etc. Built-in strategies may be devised to lessen the impact of non-critical component failures. The integrated performance of the rack and engine speed governor should ensure safe engine operation, maintain stable engine speed control, prevent undesirable engine performance (excessive smoke output, etc.) and provide responsive propulsion or power system operation.

PROPULSION/POWER SYSTEM MODEL

The overall propulsion/power system is comprised of several important subsystem models, or submodels, which are described immediately following.

DIESEL ENGINE SUBMODEL - The general approach employed in the engine model is to calculate, at any instant of time, i.e., dynamically, a single average engine torque output value as a function of the significant physical variables which affect the development of engine torque. These include engine speed, fuel quantity injected per stroke, the air-fuel ratio in the cylinder (i.e., air flow) and the thermal state of the engine. This is accomplished by means of a synthesis of: engine dynamometer data; data generated off-line from a detailed engine design analysis code set up to represent the engine under consideration [7];

and analytical equations for the dynamic variation of turbocharger speed, plenum pressures and temperatures, air and exhaust flow rates and mean in-cylinder structure temperature. More detail on this approach follows in the next several paragraphs and in the cited references.

For the purposes of evaluating the development of engine torque in response to changes in fueling and engine speed, a regression based formulation, as described and developed in [2], has been adopted as the starting point. Employing this approach, and noting that injection timing in the system under consideration is fixed by the pump characteristics and is not a separately controllable parameter, a steady-state value of brake torque may be calculated as a function only of the fuel quantity injected and the engine speed. Figure 2 shows normalized brake torque calculated from such a regression over the full operating range of the engine considered in this work, extending from 600 to 2200 RPM and from zero fueling to high levels of fueling. The regression extends into the motored negative torque region for low fueling values.

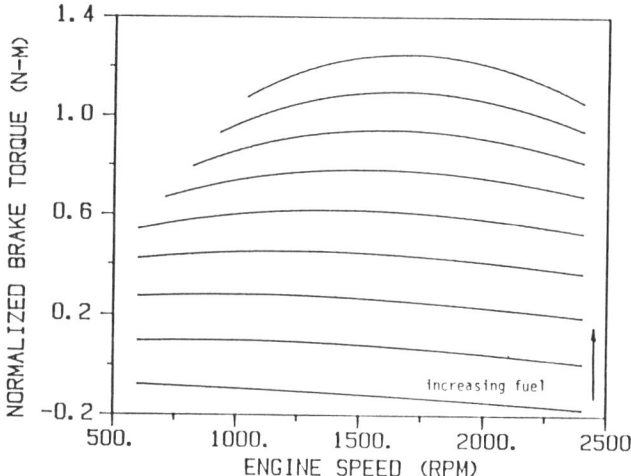

Figure 2. Normalized brake torque vs. engine speed at constant levels of fueling for a four-stroke, turbocharged and intercooled diesel engine.

Friction torque is determined from a separate regression based on motoring friction (and/or engine coast-down) data. Separate specification of friction torque allows: (1) calculation of torque values at low fueling by means of interpolation between the friction torque at zero fueling and a low positive value of torque at a suitable minimum fueling level; and (2) calculation of steady-state indicated torque, as required for correcting engine output torque for dynamic effects, described subsequently.

A regression similar to the torque regression is used to determine the fuel quantity injected as a function of rack position and engine speed, as determined from the characteristics of the pump-line-nozzle fuel delivery system. Therefore, from any set of computed dynamic values of rack position and engine speed, a steady-state value of brake torque may be obtained.

Values of engine torque computed from the regression equations correspond to fully equilibrated, steady-state conditions throughout the engine, in particular, pressures, temperatures and flow rates in the air system and temperatures within the engine structure. During a transient, instantaneous equivalence ratios and in-cylinder structure temperatures can differ substantially from the steady-state values which are implicit in the torque regression, resulting in actual torque levels which can differ significantly from their steady-state values. Torque correction factors are, therefore, applied to account for these dynamically varying combustion and heat transfer conditions. The correction factors are defined as the ratio between indicated torque at the dynamic value of equivalence ratio (or in-cylinder structure temperature) and indicated torque at the equivalent steady-state conditions. They are determined "off-line" from a detailed diesel engine design analysis code [7] and are stored within the DSM code. Instantaneous values of equivalence ratio are determined in the dynamic simulation through a separate evaluation of the diesel engine's air handling system, which also provides the instantaneous values of boost pressure required by the fuel control algorithms. The dynamic value of mean in-cylinder structure temperature is determined from the steady-state value which is used as a "driving force" in conjunction with a time constant based on the thermal inertia of the engine. The steady-state structure temperatures, over the entire engine operating envelope, are obtained "off-line" from the engine design analysis code [7].

The mathematical formulation of the engine air system is based on a combined quasi-steady component treatment and a filling and emptying analysis of the major plenums. The full characteristics of the turbocharger compressor and turbine are represented, as determined from turbomachinery maps. The treatment is general and allows easy extension to similar engine/air system configurations through modification of input data (i.e., turbomachinery, intercooler, engine geometry, etc.). Predictions are made for both steady-state and dynamic values of all pressures, temperatures and mass flow rates through the turbocharger, intercooler and engine cylinders as well as in all connecting plenums. The actual engine structure temperature, which is dynamically computed, influences the computed exhaust temperature and, hence, turbocharger response and performance. More detailed description and demonstration of these modeling approaches can be found in [2, 3] with a development of the air system governing equations given in [2].

With the instantaneous and steady-state values of equivalence ratio and in-cylinder structure temperature known, the dynamic torque corrections and, hence, the instantaneous average brake torque may be calculated. Finally, a delay of torque development with respect to a change in fueling is introduced to account for the time for fuel to burn in the first power stroke following the fueling change and the time for the new torque level to be produced in a sufficient number of cylinders so as to be assignable to the engine as a whole.

Prior to integration of the diesel engine submodel with the other system submodels, validation of the engine model is performed through comparisons of stand-alone diesel engine submodel simulations with steady-state and transient engine data. This data includes parameters such as steady-state brake torque, turbocharger speed, air mass flow rates, plenum pressures and temperatures, and transient data (i.e. turbocharger speed, boost pressure, etc.) collected during engine accelerations, resulting from fueling or load changes.

CONTROL SUBMODEL - The microprocessor-based fuel controller and governor is treated as a separate subsystem in the simulation. General control strategies have been implemented and are structured to allow easy modifications or additions. These general features include: 1) calculation of measured engine speed based on a specified number of timing pulse inputs and corresponding time increments (from the crankshaft position sensing algorithm, see below); 2) all-speed governor control with reference RPM based on throttle position and a variably specifiable droop level; 3) PID calculation of desired fuel quantity based on RPM error, with separately specifiable minimum and maximum fuel limits and, in addition, fueling limits as a function of boost pressure; 4) calculation of desired rack position as a function of measured engine speed and desired fuel quantity to be injected. Within the control system simulation, different parts of the control calculations may be carried out in asynchronous loops of separately specifiable loop times in accord with required update times. Control outputs from any loop include delays due to finite processing times.

CRANKSHAFT POSITION SENSING SUBMODEL - A generalized crankshaft position submodel, based on a discrete timing pulse sensing algorithm has been developed for the purposes of simulating the measurement of engine speed. The actual crankshaft position determination is based on sensing the location of timing teeth (with respect to a proximity sensor) on a cam which rotates once for each engine cycle (720°). Figure 3 illustrates a 12 tooth timing cam layout for a 6 cylinder, 4-stroke engine. As indicated in the figure, TDC of cylinder #1 is defined as 0° crank angle with respect to the proximity sensor. The calculated time increments between an arbitrary specifiable number of pulses are used to determine an average rate of change of crankshaft position over the duration

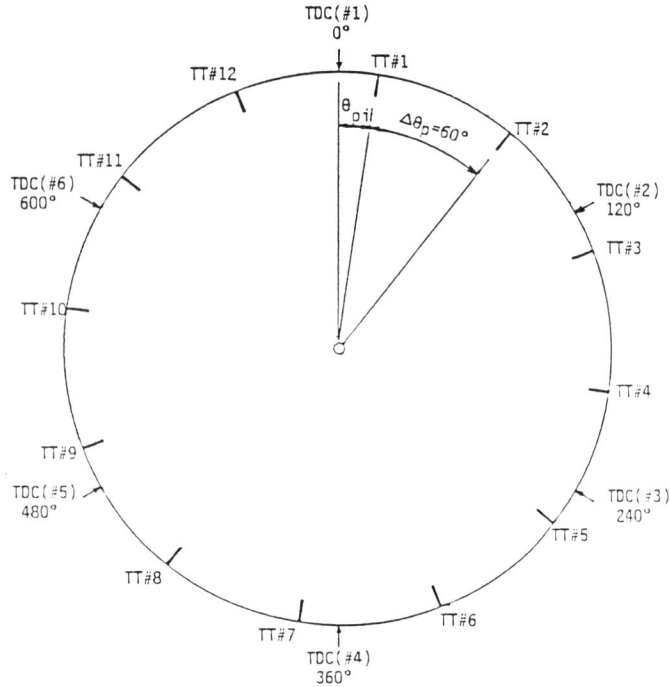

- θ_{pi} - displacement of first timing tooth from TDC, cylinder #1
- $\Delta\theta_p$ - engine crank angle interval between adjacent timing teeth
- cylinder numbers indicate position in firing order
- TT - timing teeth

Figure 3. 12-tooth timing cam layout for a 6 cylinder, 4-stroke engine, which rotates once for each engine cycle.

of the number of pulses specified, which is directly equivalent to an evaluation of engine speed.

In the simulation, the crank angle location of the timing teeth are known exactly only at points in time corresponding to simulation time steps. Linear interpolation with respect to crank angle is therefore used to determine the precise time at which a timing tooth passes the proximity sensor. In order to improve the accuracy of predicted values of measured engine speed, the simulation time increment of the timing pulse routine is chosen so that several simulation time increments elapse during the time required for consecutive timing teeth to pass the proximity sensor.

RIGID LOAD SUBMODEL - In general, simulations of the type considered here may be coupled to complex driveline or other load system submodels through the interface of torque and speed at the output shaft of the engine [2, 3, 8]. In this way, complicated driveline and driveline-engine-controller interactions can be studied. For the purposes of evaluating the design and calibration of the rack actuator and governor control elements of the system, which may ultimately perform in a variety of different load system applications, a simple rigid load submodel variably geared to the engine has been constructed.

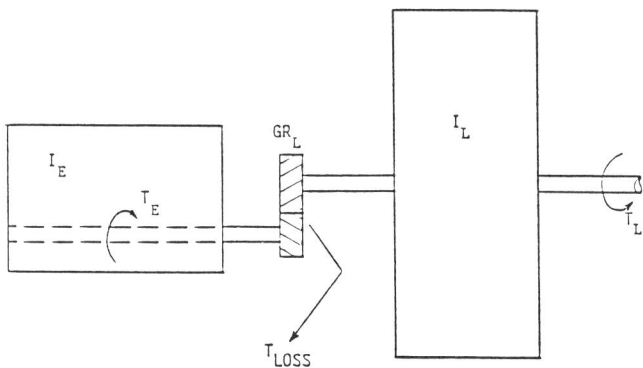

Figure 4. Schematic of the variably geared, rigid load submodel.

A schematic of the rigid load subsystem is shown in Figure 4. The system is characterized by separately specifiable engine and load inertia (I_E and I_L); a gear ratio between the engine and load (GR_L); load torque applied to the rigid load (T_L); and a torque loss in the gear set (T_{LOSS}). The torque loss is represented as a linear function of input torque, starting with a no-load churning loss at a zero value of input torque to the gear set. In this case where the interconnecting shafts are non-compliant, engine speed is directly proportional to load speed, and equivalent values of load torque and total system inertia, as seen at the engine, may be determined in a straightforward manner.

RACK ACTUATOR SUBMODEL - This section contains a description of the rack actuator and its control, as well as simulations demonstrating its response under different conditions. Figure 5 shows a schematic of the rack actuator mechanical and electromagnetic components which are included in the rack actuator submodel. The rack actuator is treated as a second order, spring-mass, linearly damped system, with external force applied through a linear magnetic solenoid. The mathematical formulation is standard, i.e., ordinary differential equations for rack displacement and velocity, and will not be presented here.

The core is an iron plunger, which becomes magnetized by the field of the solenoid. When magnetically saturated, the core is subjected to a force which is proportional to solenoid current and varies according to the distance between the magnetic centers of the plunger and solenoid. Prior to being saturated by the field, the force on the plunger varies as the current squared [9]. Solenoid force as a function of rack position and solenoid current, which has been idealized from data in [9], is shown in Figure 6. A maximum allowable level of solenoid current is implicit in Figure 6, which in practice would be limited by the rack actuator's current controller. Also shown in the figure is the magnitude of the spring force as a function of rack position. Positive and negative values of the difference between linear magnetic solenoid force and return spring force produce acceleration and deceleration of the rack, respectively. In this figure, the spring constant and pre-load of the spring have been chosen so as to produce nearly the same amount of available force to accelerate the rack at low rack values, as is available to decelerate the rack at high rack values.

Control of rack position is accomplished by varying solenoid current in response to changes in the deviation between actual rack position and desired rack position. In general, the solenoid current will have a finite rise-time, dependent on the ratio of the solenoid resistance and inductance, and, in addition, will require separate control through high frequency variation of applied solenoid voltage. However, in the present applications we will

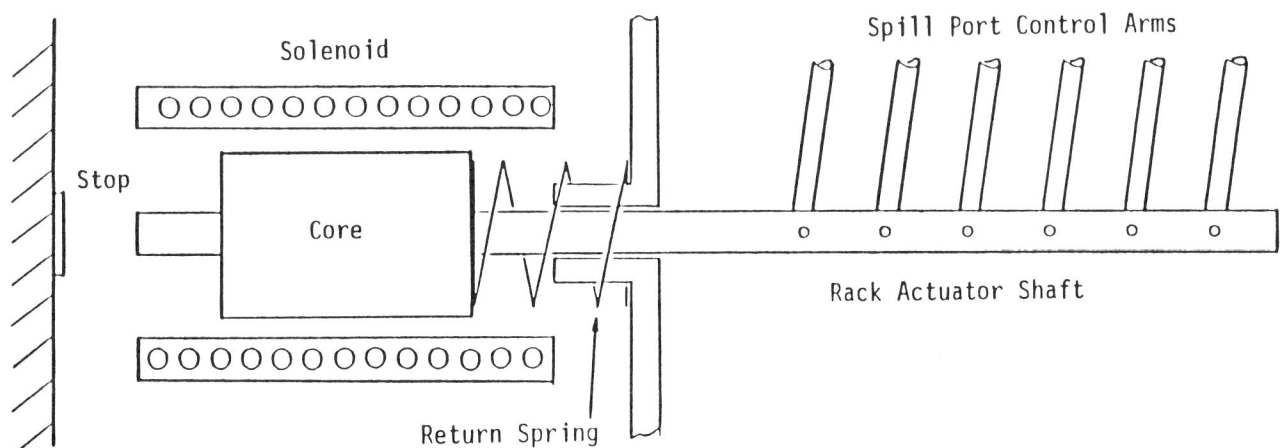

Figure 5. Schematic of the rack actuator mechanical and electromagnetic components.

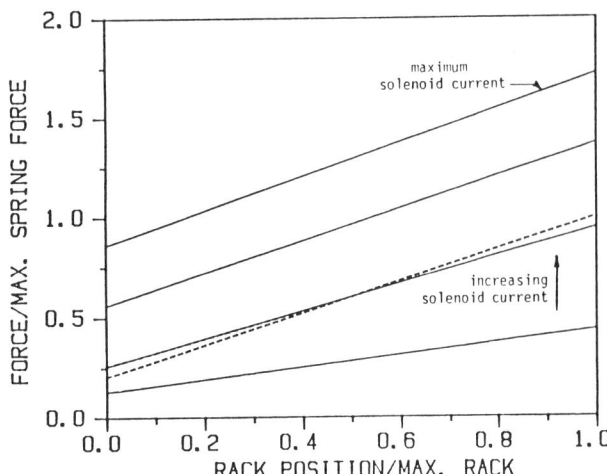

Figure 6. Normalized spring force (dashed line) and linear magnetic solenoid force at constant levels of solenoid current (solid lines).

only consider the case of small solenoid time constants and fast, accurate solenoid current control so that solenoid current may be assumed to be equal to its desired value at all times. Suitable sensing of rack position by means of a displacement transducer provides feedback to the rack controller. The rack control logic is assumed to be implemented in stand-alone analog circuity, such that rack control is exercised continuously both in practice and in the simulation. Note that within the framework of DSM, digital control of rack could easily be simulated as well, implemented in either a separate microprocessor or integrated with the fuel control/governor microprocessor. The analog control simulation in this work would correspond quite closely to a stand-alone digital control with rapid update time (< 5 msec).

Open Loop Rack Control - With this type of actuator, production designs of rack control will be based on closed loop methods in order to realize high accuracy and excellent dynamic response [6, 10]. DSM, however, lends itself easily to an evaluation of alternative strategies and, for the purposes of the current investigation, allows comparisons of closed and open loop performance. In the event of rack position sensor failure, it would be desirable to have a secondary or backup mode of rack control based on open loop control until suitable repair could be made [11].

Open loop control of rack position is achieved in the simulation by setting solenoid current, at any instant of time, to its required value at the desired rack position (see Figure 6). These values of solenoid current correspond to intersections of the solenoid force and the return spring force curve as a function of rack position. Note that with this type of control, the forces available to accelerate and decelerate the rack mass are dependent on the relative magnitude of the spring constant, k_{SPR}, and the slope of the linear magnetic solenoid force curves, k_{LMS}. Preferred designs of open loop control would maximize k_{SPR}/k_{LMS} in order to increase this available force, in conjunction with increased mechanical damping to avoid over/undershoot. A k_{SPR}/k_{LMS} ratio less than one produces an intrinsically unstable system, since any movement of the rack from an equilibrium position will produce a resulting force in the direction of motion. As will be demonstrated, the performance of the closed loop controller is relatively insensitive to this open loop stability constraint.

Closed Loop Rack Control - Closed loop control of rack position is achieved through control commands for changing solenoid current in response to rack position error (desired minus actual position), as determined from the position feedback sensor. A PID algorithm with constant gains has been implemented for this purpose but may be easily modified to evaluate alternative strategies (i.e., functionally dependent gains, etc.). The calculated PID value of solenoid current may be limited to a specified maximum value. Note that the closed loop controller derives its responsiveness from the availability of instantaneous maximum and minimum linear magnetic solenoid force for any given rack position, rather than being limited to the final desired value of force as in the open loop case. In addition, the closed loop control protects against long term drift in solenoid or spring rate characteristics, both of which could negatively impact the accuracy of open loop control.

Simulated Rack Response - With the given rack control and actuator descriptions, the response of rack may be investigated through externally applied commands for changing desired rack position. In this way, using DSM, the response of the rack may be evaluated in a stand-alone mode, prior to integration with the other propulsion/power system submodels.

The response of the open loop controller to a square pulse in desired position for three different values of mechanical damping are shown in Figure 7. Here, rack position and desired rack position are shown normalized with respect to the maximum rack displacement. Requested values of desired position are for over 50% of the allowable rack travel and span the maximum and minimum fueling locations with respect to the overall speed range of the engine. The values of damping ratio indicated in the figure are calculated based on k_{SPR}. Note that critical damping occurs at a damping ratio (ξ) less than 0.5, since the system sees an equivalent spring constant less than k_{SPR} (i.e., $k_{SPR} - k_{LMS}$). Similarly, the frequency of oscillations are smaller than the (natural) frequency of the spring-mass system, in response to a step input in external force. The frequency of oscillation is seen to be slightly dependent on rack position, due to the characteristics of the linear magnetic solenoid force. Damping ratios less than 0.3 produce

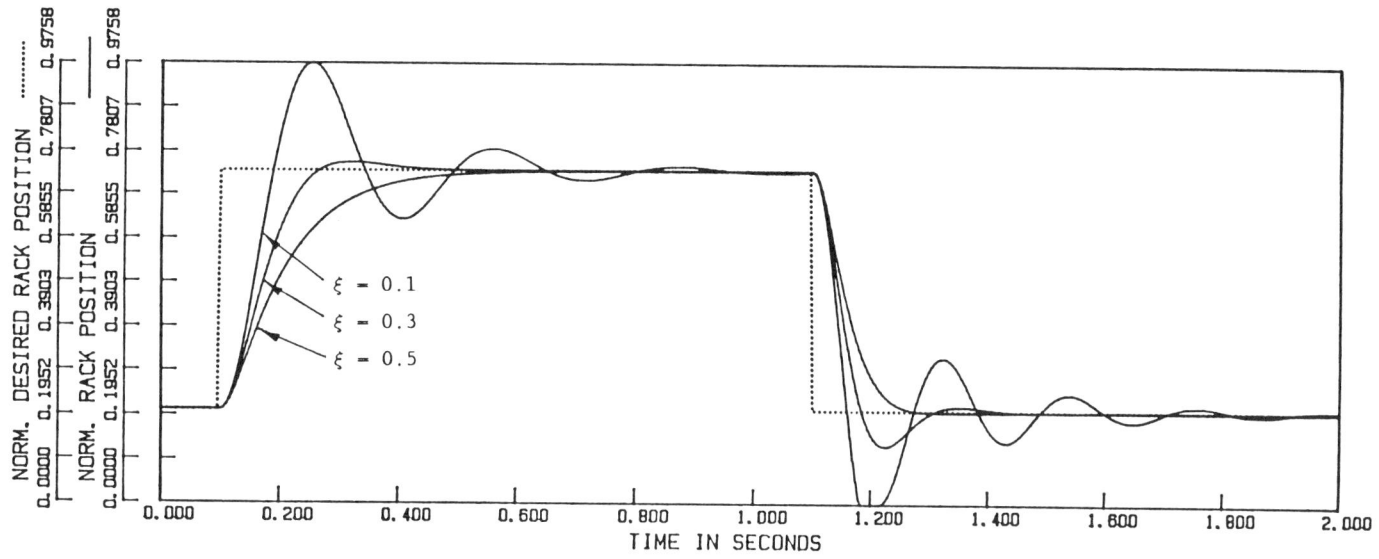

Figure 7. Response of open loop rack control to a square pulse in desired rack position as a function of damping ratio.

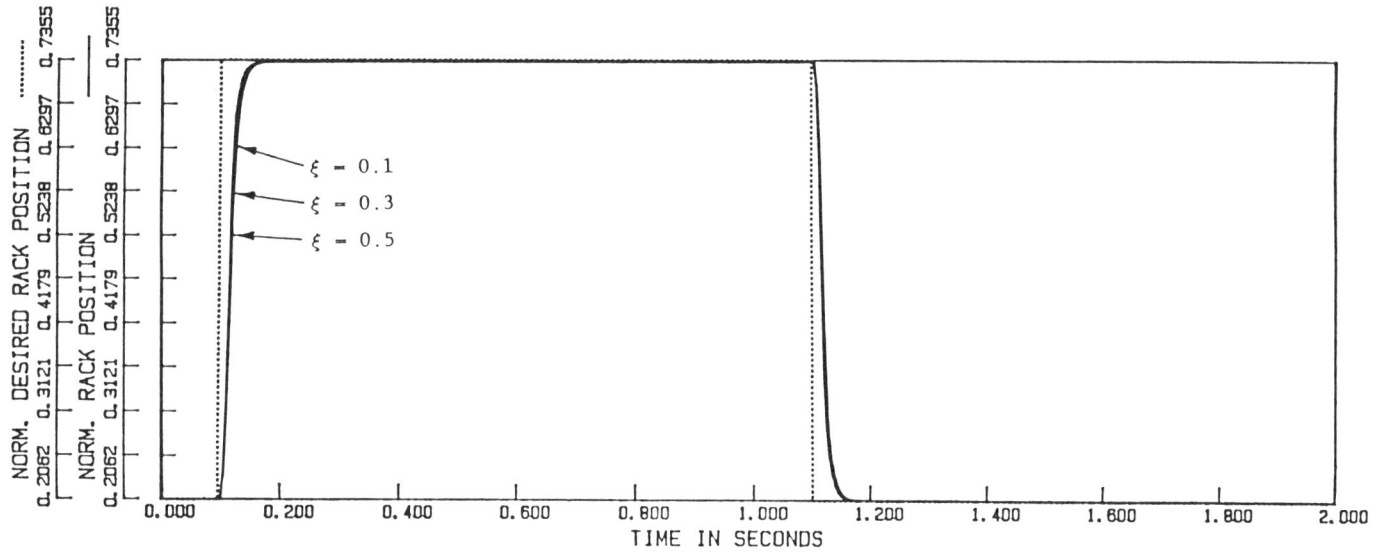

Figure 8. Optimized response of closed loop rack control to a square pulse in desired position as a function of damping ratio.

overshoot and oscillation of rack position. As indicated in the figure, a damping ratio of 0.1, although reaching the desired level rapidly, produces large overshoot and oscillation and causes the rack to bump up against its lower stop at approximately 1.2 seconds. For the purposes of a backup open loop control mode, the rack response of $\xi = 0.5$ could be considered acceptable. The results of Figure 7 indicate that rack designs incorporating this type of control would require attaining and maintaining a specific level of mechanical damping, which could be difficult in practice. Further evaluation of the open loop rack controller in the context of the overall propulsion/power system is presented below (see Simulated Propulsion/Power System Response).

The response of the closed loop controller and rack actuator to the equivalent change in desired rack position is shown in Figure 8. Results were obtained for the same damping ratios as the open loop response, and fixed PID gains which produced the optimal response indicated. Unlike the open loop control, mechanical damping (in the range considered) has only a minor effect on the response of the rack. This will be true in general if the magnitude of the damping force is much smaller than the maximum force available to accelerate or decelerate the rack. This, in turn, is dependent on the physical constants of the system. The optimal gains provide the desired change in position in less than 50 milliseconds, with no over- or undershoot. This is in comparison to the open loop controller which required over 250 milliseconds for the critically damped case (Figure 7).

Additional studies of the closed loop controller were performed with varying spring constants. A significant result of these studies is that a response near that of Figure 8 could be obtained for perturbations of k_{SPR} about k_{LMS} ($\pm 10\%$), through modification of PID gains. Although not practical for designs which consider incorporation of the open loop control discussed above, due to intrinsic instability, this result was also true for the case of $k_{SPR} < k_{LMS}$.

SIMULATED PROPULSION/POWER SYSTEM RESPONSE

Subsequent to rack actuation and control design and analysis, engine speed governor gains may be calibrated and rack performance evaluated in the context of the overall propulsion/power system. This is accomplished in the simulation through the use of the integrated propulsion/power system simulation code.

STEPPED THROTTLE AT CONSTANT LOAD - A simulation featuring the response of the rack in combination with an all-speed engine governor to a sequential step increase and decrease in throttle position is indicated in Figures 9a through 9d. In Figure 9a all speeds are shown normalized with respect to rated engine speed. This simulation is specified with a free engine (external load inertia = 0) and a small and constant external load torque. Droop is set to zero (i.e., isochronous control) so that a step increase in throttle at t = 0.1 seconds from near one quarter to full throttle produces a step change in the desired value of engine speed (engine speed set point) to the rated speed. Governor gains have been optimized to produce the near optimal response shown here.

Engine speed climbs in a linear manner as rack and fueling (Figures 9b and 9c) respond to the underspeed condition. Note that measured engine speed tracks actual (dynamically predicted) engine speed extremely closely for the rate of change of speed shown. Some minor ringing in the rack is indicated in Figure 9b as the engine approaches its low speed idle condition. This, however, does not appreciably affect engine speed (Figures 9a, t ~ 9.0 seconds). Fueling is limited by available boost pressure during the engine acceleration as indicated by Figure 9c which shows fueling equal to only fifty percent of the fuel quantity at maximum torque (i.e., torque curve fuel value) that could be supplied at rated speed. Figure 9d shows the engine brake and load torque, respectively. As indicated, engine acceleration and deceleration are a result of positive and negative differences between brake and load torques, respectively. Evident is the higher acceleration rate of the engine compared to the deceleration rate, which is due to the relationship between available engine torque, friction torque and load torque.

This simulation sequence was also performed for the open loop rack control demonstrated in Figure 7 ($\xi = 0.5$). The engine speed and rack response for this simulation are indicated in Figures 10a and 10b. As indicated, the large lag between desired and actual rack position results in appreciable ringing of rack and

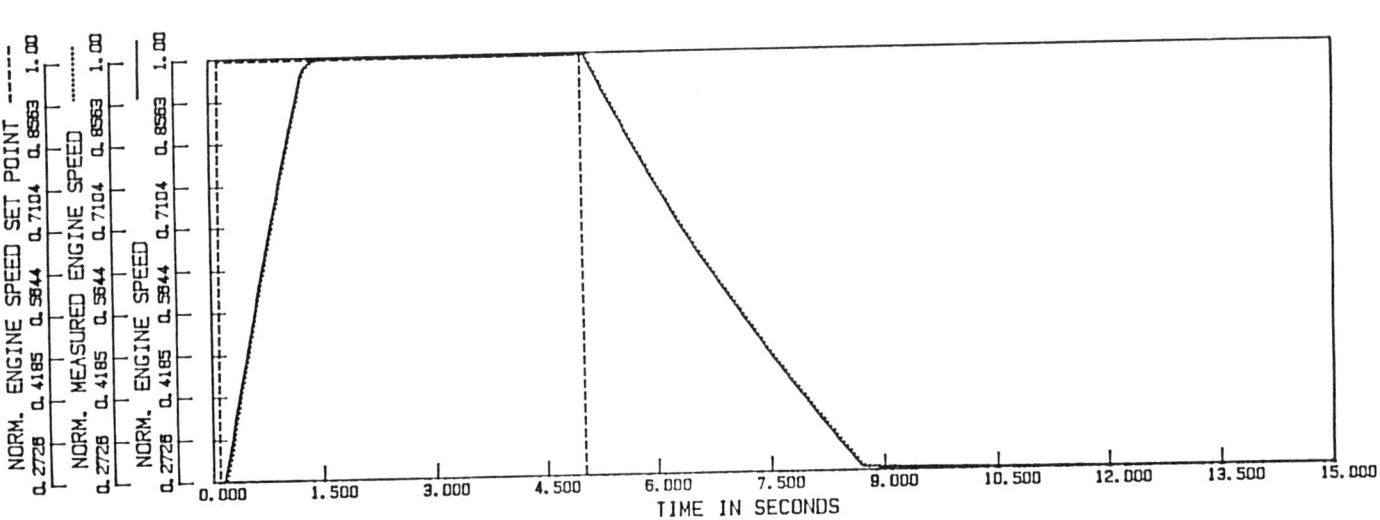

Figure 9a. Engine speed, measured engine speed and engine speed set point for a step increase and decrease in throttle position at constant load with closed loop rack control.

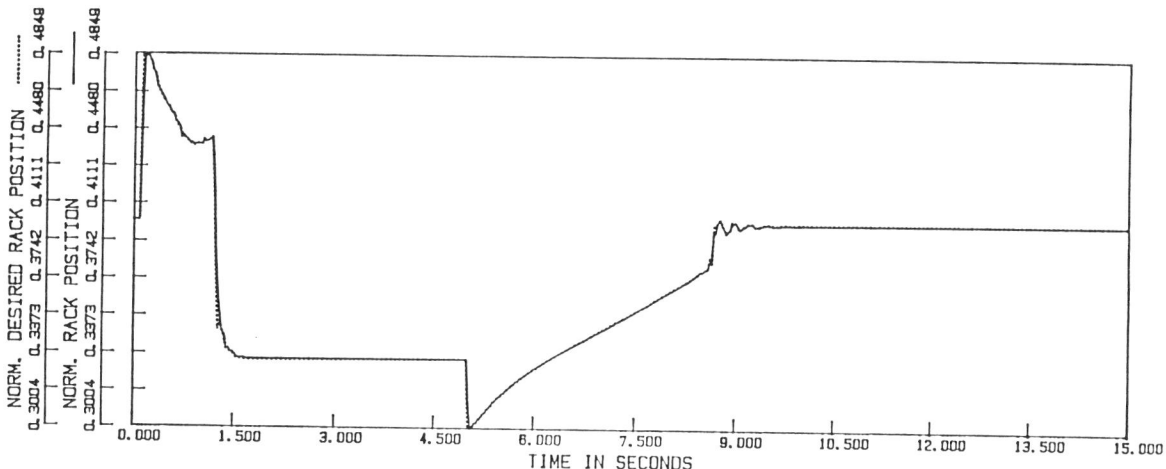

Figure 9b. Response of rack and desired rack position for a step increase and decrease in throttle position at constant load with closed loop rack control.

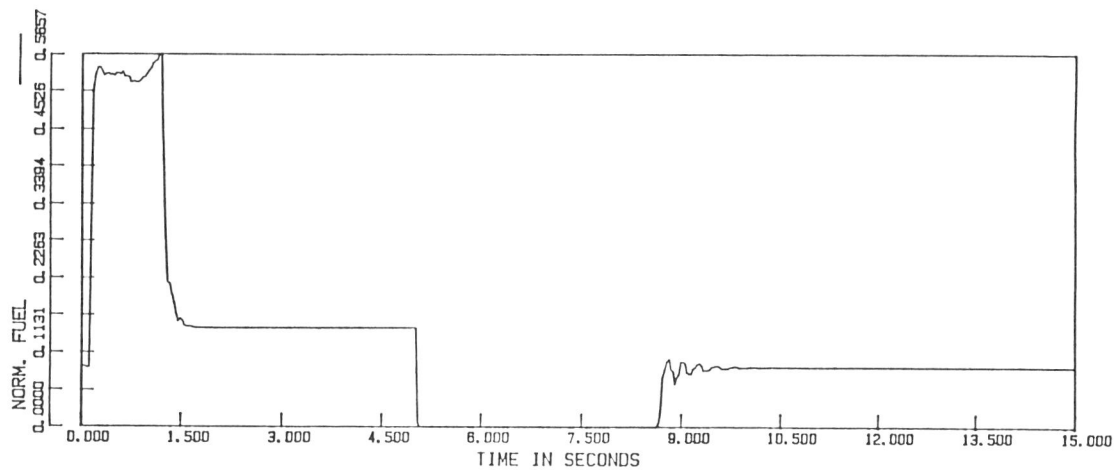

Figure 9c. Engine fueling for stepped increase and decrease in throttle position at constant load with closed loop rack control.

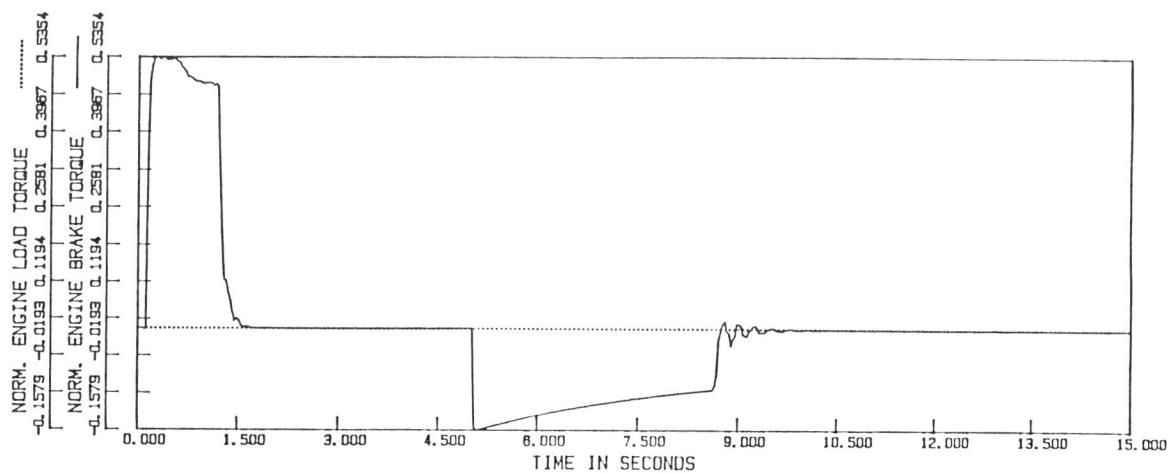

Figure 9d. Engine brake and load torque for a stepped increase and decrease in throttle position with closed loop rack control.

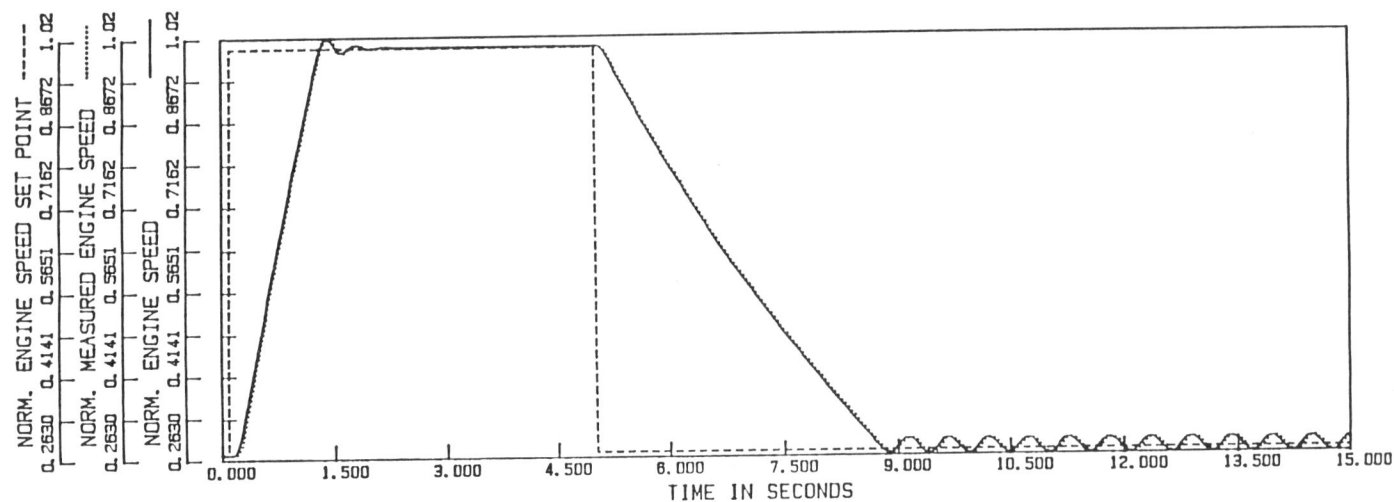

Figure 10a. Engine speed, measured engine speed and engine speed set point for a step increase and decrease in throttle position at constant load with open loop rack control.

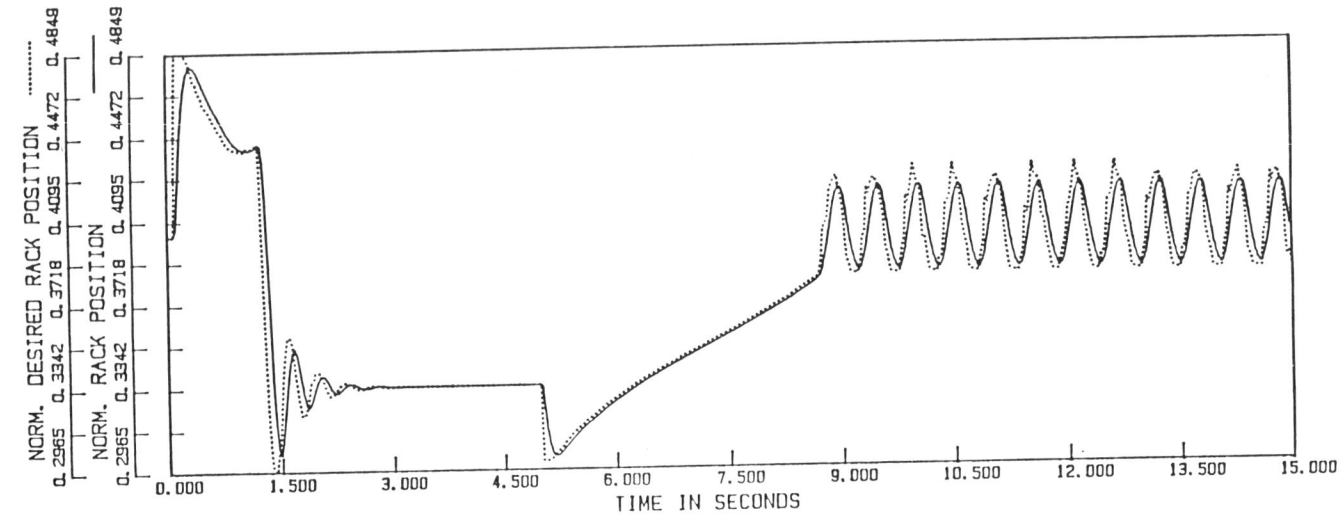

Figure 10b. Response of rack and desired rack position for a step increase and decrease in throttle position at constant load with open loop rack control.

engine speed at both high and low speed idle conditions. A reduction of governor gains (not shown) produces increased stability at the expense of less responsive speed control. The backup rack control strategy would have to include modification of governor gains in conjunction with the transition from closed to open loop control to obtain a more stable response. In this instance, in the event of rack position sensor failure, the open loop rack control might be considered acceptable as a secondary mode of control with limited capability.

STEPPED EXTERNAL LOAD TORQUE AT CONSTANT SPEED - Figures 11a and 11b compare the rack and engine speed responses to a change in external load torque at constant throttle for the closed and open loop rack control strategies (both with $\xi = 0.5$). This simulation is performed at full throttle at an initial level of external load torque, which is set equal to one quarter of rated load at rated speed. Load torque is then stepped to three quarters of rated load at t = 0.1s. In this case, the external load inertia is equal to five times that of the free engine. In this application, the specified load inertia is equivalent to a 20,000 lb medium-duty truck with an overall transmission and axle reduction ratio of 16.0, or, in a power system application, a typical value of generator inertia for direct drive gen-set installations. Droop is again specified equal to zero, i.e., isochronous

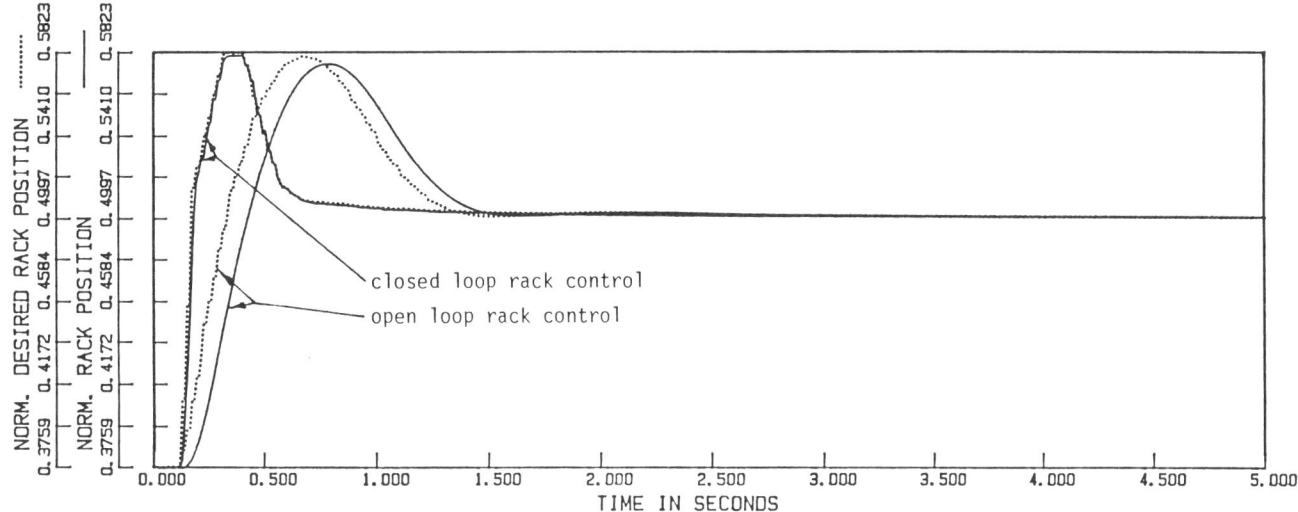

Figure 11a. Response of rack and desired rack position for a step in external load torque at constant throttle. Both the responses of a closed and open loop rack controller are indicated.

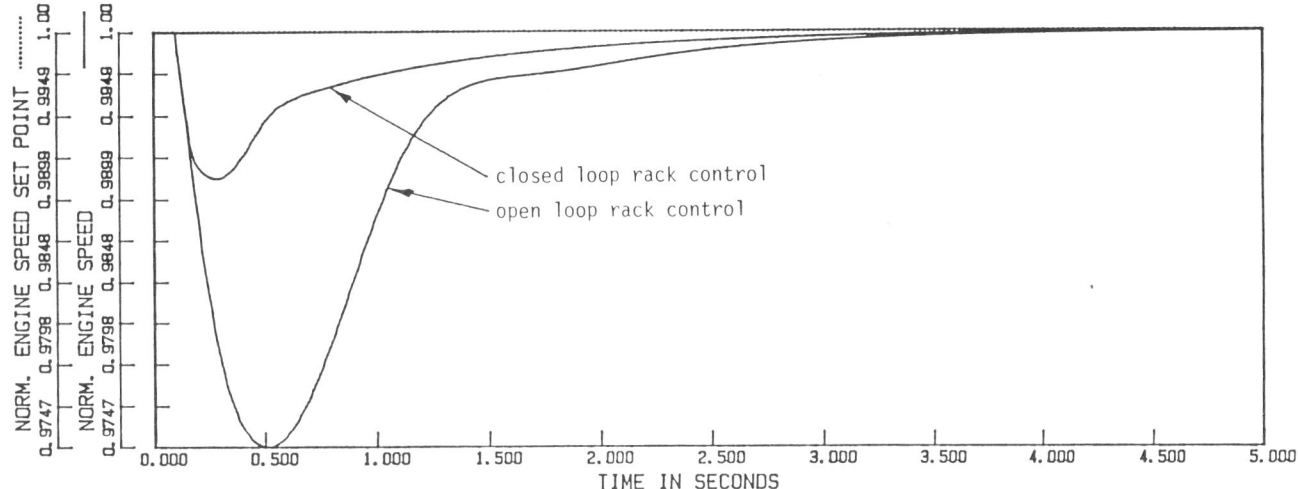

Figure 11b. Response of engine speed and engine speed set point for a step in external load torque at constant throttle. Both the responses for closed and open loop rack control are indicated.

engine speed control.

In the rack and speed responses indicated in Figures 11a and 11b, governor gains have been optimized for both the open and closed loop rack controllers. As in the previous simulation, an attempt to use governor gains developed for closed loop rack with open loop control produced unacceptable ringing in the rack and engine speed response. As indicated in both of the speed responses of Figure 11b, engine speed drops in response to increasing load torque. Differences in the responsiveness of the rack and speed governor produce relatively large differences in the magnitude of the underspeed (Figure 11b, 1% compared to 2.5% of set speed).

Here, depending on the installation in which the simulated load torques are encountered (i.e. change in road grade for the medium duty truck or a change in electrical power requirements for a diesel engine generator set), the speed response obtained with backup rack control may or may not be considered acceptable. In the latter case, alternate actuator designs and control strategies may be conceived and evaluated.

CONCLUSIONS

The utility of applying a comprehensive, non-linear dynamic simulation of a diesel propulsion/power system to the design and integration of subsystem components and control strategies has been demonstrated. In particular, the interactive performance of a solenoid-actuated electronically controlled rack and a microprocessor-based fuel controller and governor have been examined in the context of the overall propulsion and power system response. The simulation allows dynamic, quantitative analysis of complex subsystem interactions over a wide range of realistic operating conditions. In particular, it has been shown that closed loop control is far superior to open loop control of the solenoid-actuated rack, although the open loop control can serve as an acceptable backup mode of control with restricted capability. Further it has been shown that the quality of rack control has a direct impact on engine speed governing and that DSM provides a quantitative tool to aid in the development of optimized control for the entire system.

REFERENCES

1. M. J. Jennings, P. N. Blumberg and R. W. Amann, "A Dynamic Simulation of the Detroit Diesel Electronic Control System for Evaluating Engine-Driveline Interactions", 1986 Joint Automatic Control Conference, Seattle, WA.

2. M. J. Jennings, P. N. Blumberg and R. W. Amann, "A Dynamic Simulation of the Detroit Diesel Electronic Control System in Heavy Duty Truck Powertrains", SAE Paper 861959, Truck and Bus Meeting and Exposition, King of Prussia, PA, November, 1986.

3. C. I. Rackmil, P. N. Blumberg, D. A. Becker, R. R. Schuller and D. C. Garvey, "A Dynamic Model of a Locomotive Diesel Engine and Electrohydraulic Governor", Journal of Engineering for Gas Turbines and Power, Vol. 110, pp. 405-414, July, 1988 (also ASME Paper 88-ICE-26).

4. P. N. Blumberg, C. I. Rackmil and D. A. Becker, "A Dynamic Simulation of a Diesel-Electric Locomotive Propulsion System", 1988 Joint ASME/IEEE Railroad Conference, Pittsburgh, PA.

5. M. Nishimura, N. Fujitani and M. Iwatuki, "The Nippondenso In-Line Injection Pumps with Electronic Control for the Clean Diesel Engine", SAE Paper 870436, International SAE Congress and Exposition, Detroit, MI, February, 1987.

6. U. Flaig and A. Sieber, "Electronic Control Units of Bosch EDC Systems", SAE Paper 880185, International SAE Congress and Exposition, Detroit, MI, February, 1988.

7. T. Morel, R. Keribar and P. N. Blumberg, "A New Approach to Integrating Engine Performance and Component Design Analysis Through Simulation", SAE Paper 880131, International SAE Congress and Exposition, Detroit, MI, February, 1988.

8. M. C. Tsangarides and W. E. Tobler, "Dynamic Behavior of a Torque Convertor with Centrifugal Bypass Clutch", SAE Paper 850461, International SAE Congress and Exposition, Detroit, MI, 1985.

9. W. A. LaPierre, "Electrical Engineering", Mark's Standard Handbook for Mechanical Engineers, Eighth Edition, © McGraw-Hill, Inc., 1978, Section 15, pp. 77-79.

10. F. Murayama, Y. Tanaka and S. Ito, "The Nippondenso Electronic Control System for the Diesel Engine", SAE Paper 880489, International SAE Congress and Exposition, Detroit, MI 1988.

11. G. Stumpp and H. Kull, "Strategy for Fail-Safe Electronic Diesel Control System for Passenger Cars", SAE Paper 830527, 1983.

890395

Optimization of Heavy-Duty Diesel Engine Transient Emissions by Advanced Control of a Variable Geometry Turbocharger

A.D. Pilley, A.D. Noble, A.J. Beaumont, J.R. Needham, and B.C. Porter
Ricardo Consulting Engineers Ltd.

ABSTRACT

Ricardo have developed a systematic approach for the design of transient engine control strategies using advanced control techniques. The methodology was initially applied to the design of a testbed speed and torque controller. This enabled complex transient tests to be carried out with equipment normally used for steady-state testing. The same techniques were applied to the design of a controller for a variable geometry turbocharger aimed at vehicle applications. The influence of different control strategies on emissions and fuel economy was evaluated on a heavy-duty diesel engine over a section of the US FTP cycle. Particulate reductions of up to 34% were achieved without increasing NOx.

CONSIDERABLE EFFORT WILL be required for the heavy-duty diesel engine to meet the stringent US 1994 emissions limits of 5 g/bhp.h NOx (oxides of nitrogen) and 0.1 g/bhp.h particulates. Coordinated development of fuel injection equipment (FIE), combustion chamber design, air management and low oil consumption technology will be vital if 1994 levels are to be approached. Ultimately 'engine-out' emission levels might not be reduced to 1994 levels and exhaust aftertreatment may be necessary. Cost effective, reliable trap and catalyst systems do not currently exist, so the engine manufacturer must work on solutions to 'engine-out' emissions levels in parallel with continued investigation of exhaust aftertreatment.
Over the past decade turbochargers with a variable geometry turbine stage have received serious consideration from a number of manufacturers (1-6)*. Under steady-state conditions the application of variable geometry turbochargers (VGT) has produced significant increases in torque back-up and some useful improvements in fuel consumption and exhaust emissions (7, 8, 9).

* Numbers in parentheses designate references at end of paper

Most of the control strategies developed for VGT have been designed to increase boost pressure to improve air/fuel ratio and/or increase low speed torque. For vehicle applications, a finite number of turbine areas (usually 3 or 4) are typically used to cover the engine speed and load range. The simplest form of control allows the turbine area to increase with engine speed (10). At light loads, reduction in turbine area at low speeds will adversely affect fuel consumption and so some form of load control has also been incorporated, for example using the accelerator pedal position as a load sensor (4). Feedback of boost or exhaust pressure is commonly used to regulate turbine area in closed loop control systems (11).
Little work has been published to show the influence of variable geometry turbocharging on exhaust emissions other than smoke (12, 13). At high load conditions smoke and particulates may be reduced by employing higher air/fuel ratios. Similarly hydrocarbon (HC) emissions can be reduced during cold starting by increasing exhaust back pressure (increasing engine load) and at high-speed light-load conditions by increasing boost pressure and temperature to reduce ignition delay (9).
Control strategies for transient operation have, in the main, been developed from steady-state test data (5, 12, 13). These have been adapted to incorporate some form of 'anticipation' to account for transient conditions, for example based on rack position history (5).
Some work describing a systematic approach for transient engine controller development has been reported. Winterbone and Jai-In (14) used identification techniques to derive dynamic models for a turbocharged diesel engine with a VGT. Experimental measurements were used to validate a non-linear 'filling and emptying' simulation that was to be used in the design of a multivariable controller. Linear first order models were also derived for the actual controller but no results were reported.

VGT control strategies must be developed to improve transient engine performance (driveability) and emissions. Current steady-state development methods will therefore have to be extended to include a transient development phase. However the cost of purpose-built transient test facilities is high and their numbers are limited.

As part of an internal research programme, Ricardo have developed techniques for transient controller optimisation using existing steady-state testbeds. Methods have been developed for controlling the engine during complex transients using a standard eddy-current dynamometer and excitation unit. The same techniques have been applied to the design of a controller for a VGT.

This paper describes the systematic approach that has been used in the design of the testbed speed and torque controller and the VGT controller. The advantages of this approach are illustrated through the design and evaluation of transient control strategies for a VGT. The same methodology can be used in the design of future on-board engine controllers.

OBJECTIVES

The work described in this paper is part of an integrated research and development programme with the principal objective of developing a heavy duty diesel engine that will achieve the US 1994 emissions targets.

The objective of this stage of the work was to develop techniques to assess improvements in exhaust emissions through the use of optimised transient control strategies for a VGT.

The specific objective for the VGT controller was to minimise particulate emissions, without increasing NOx, over a section of the US FTP heavy-duty diesel transient test cycle.

CONTROLLER DEVELOPMENT PROCEDURE

Modern control techniques have been widely used in the aerospace industry for a number of years. Interest has also developed in these more systematic approaches for the design of controllers for I.C. engine applications - most notably for spark ignition engine management systems.

Modern control techniques have three important advantages over traditional classical methods:

1. The controller is multivariable, ie. it can simultaneously control a number of parameters in a coordinated manner. Classical controllers, such as Proportional-Integral-Derivative (PID) controllers, can generally only handle systems with single inputs and outputs.

2. The controller incorporates dynamic models of the system. It can therefore predict how the system will respond to changes in input parameters (and measureable disturbances) before deciding the appropriate action. Classical controllers generally only react to changes based on the measurement of events, ie. they react after an event has occurred.

3. The controller results directly from the minimisation of a mathematical function that expresses the objectives for the system, eg. to minimise fuel consumption and/or particulate emissions. This allows a more systematic design than classical techniques.

The controller design process used by Ricardo can be divided into six principal phases:

1. System analysis.
2. System identification and modelling.
3. Model validation.
4. Control system design.
5. Simulation.
6. Controller implementation.

The following sections give a brief overview of the approach, further details are given in (15).

SYSTEM ANALYSIS - An initial analysis is first performed to define the scope of the problem. The controller objectives are defined and the system inputs and outputs are categorised into:

(a) controllable inputs,
(b) uncontrollable inputs (external disturbances),
(c) measurable outputs.

Figure 1 illustrates the engine and testbed system used in this work and indicates the parameters being controlled and measured by both the speed and torque controller and the VGT controller.

SYSTEM IDENTIFICATION AND MODELLING - The engine system (including the dynamometer, fuel pump, governor, VGT and associated actuators and sensors) is considered as a 'black box' with controllable inputs (speed demand, brake torque demand and turbine area demand) and measurable outputs (governor demand lever position, engine speed, brake torque, rack position, VGT guide vane position (GVP), boost pressure, exhaust pressure, turbocharger speed, airflow, HC, NOx, smoke, particulates, etc).

Experimental data is obtained by simultaneously applying uncorrelated signal perturbations to the inputs and measuring the response of the engine. The perturbations are generated using a pseudo random binary sequence (PRBS) which enables the maximum amount of dynamic information to be obtained for a given test length and signal amplitude.

Due to the non-linearities in the behaviour of turbocharged diesel engines, identification tests are performed about different nominal speeds and loads. There will also be significant time delays and distortion of some

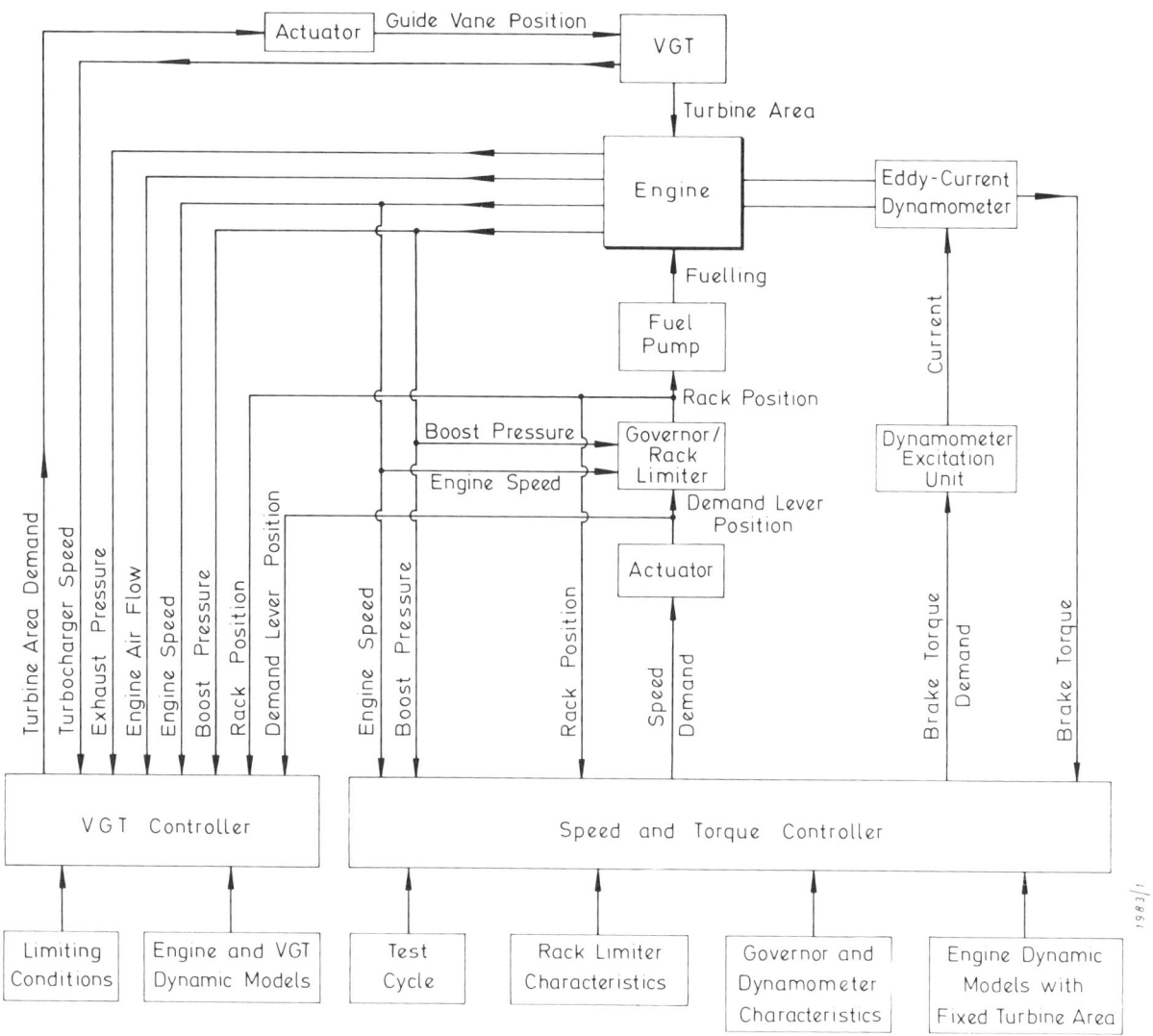

Figure 1. Testbed/Controller Configuration.

of the engine data due to the nature of the measuring systems. The response characteristics of the measuring equipment are therefore included in the identification analysis. Emissions analyser signals, for example, have to be 'reconstructed' to compensate for the transport delays and analyser response characteristics (16).

Mathematical models are 'identified' that allow the outputs to be predicted from the inputs, together with auxiliary variables that can be measured and/or calculated. A large number of parameters are measured during the identification phase, but only a limited number of these will be required for the final controller. Transfer functions relating each of the outputs (eg. engine speed, torque, smoke, HC, NOx, etc.) to the inputs (speed demand, brake torque demand, turbine area demand) are calculated. The system is represented in discrete time by state space matrices. Measurement noise is also characterised in the models.

It is worth emphasising that the models are developed to enable the controller to predict the behaviour of parameters that cannot normally be measured in a practical application (eg. smoke, HC, particulates). This approach implicitly assumes that the relationships between input and output parameter changes will remain valid over the life of the controller.

MODEL VALIDATION - The dynamic models are validated by comparing the response of the model with that of the engine when subjected to the same input parameter changes. This produces a set of validated linear models describing the dynamics of the engine at a variety of operating conditions.

Gain scheduling techniques can be used to derive the appropriate engine model as the engine moves through its operating range if the non-linearities are significant. Alternatively the non-linearities can be mapped and included in the identification analysis.

CONTROL SYSTEM DESIGN - The validated dynamic engine models are then used in the design of control laws. These are formulated by specifying some performance objective for the control system. The 'cost function' is a mathematical expression that might contain, for example, the conflicting objectives of reducing exhaust emissions and fuel consumption. The control law is designed using an advanced technique, in this case Generalised Predictive Control (GPC), see (17) for details. Having identified the engine models and established the GPC design parameters the controller is calculated automatically.

For the case of the testbed controller, which was designed to operate the engine over pre-defined test schedules, 'preview' of future demands can be incorporated into the controller design. A control strategy that has knowledge of future engine demands is not considered realistic for practical in-vehicle applications, such as a VGT controller.

SIMULATION - Simulation of the engine and control system enables different control strategies to be assessed in a rapid and cost-effective manner. The use of simplified control strategies for implementation in production engine management systems can be simulated to establish the requirements for the final control system. At this stage it is possible to estimate the expected improvements in the performance of the engine in terms of fuel economy and exhaust emissions and so decide the value of further test work.

CONTROLLER IMPLEMENTATION - The control strategy would normally be tested and developed using an 'off-the-shelf' computer, such as an IBM PC. The controller performance can be verified by comparing the behaviour of the system with that predicted by the simulation. Some tuning of the GPC weighting parameters may be required. Having designed the control strategies, the requirements for a production system could be defined at this stage.

TEST PROGRAMME

TEST SCHEDULE SELECTION - The US FTP heavy-duty transient cycle includes substantial periods of idle and near steady-state operation. There are seven major transient sections of varying complexity and duration.

Consideration of the transient speed and load changes and NOx/particulate contribution of various parts of the cycle led to the selection of the second part of the New York Non-Freeway (NYNF) cycle for test purposes, see Figure 2. The NYNF[2] section (from time 190 to 262 s) consists of a rapid acceleration and deceleration and has severe rapid load changes. Measurements on typical contemporary turbocharged, aftercooled diesel engines indicate that this section has the highest mean particulate rate of any of the truly transient sections of the FTP cycle. This section appears twice in the FTP test and typically contributes around 15-20% of the total particulates and 10-15% of the total HC and NOx emissions during the 'hot' cycle. It is therefore ideal for

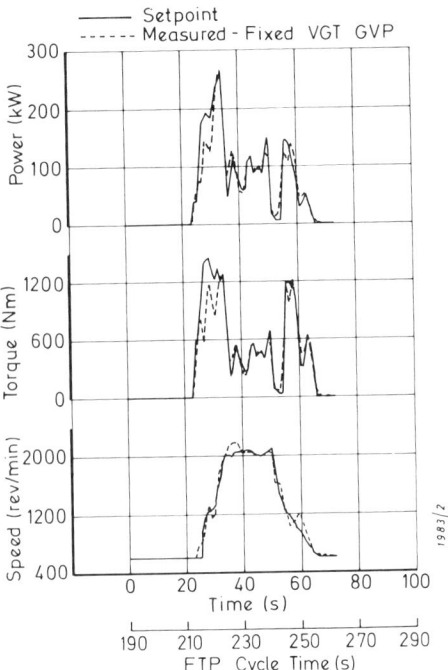

Figure 2. Comparison of Setpoint and Measured Speed, Torque and Power over the NYNF[2] Cycle with Fixed VGT Guide Vane Position.

demonstrating the effectiveness of the methodology and the development of low emissions VGT control strategies.

ENGINE DETAILS - The experimental test work was carried out using a Volvo TD120F engine, details of which are given in Appendix A. This engine was developed in a previous Ricardo research programme to improve torque back-up, fuel consumption and emissions through the use of rematched FIE and VGT, see (7) for details. With a fixed geometry turbocharger this engine is typical of modern European turbocharged, aftercooled diesels and, with retarded injection timing, produces emissions levels consistent with a US 1990 engine over the NYNF[2] section of the FTP cycle.

Development of better oil control, increased compression ratio, rematched high pressure FIE and combustion system design would be required for this engine to satisfy 1991 emissions levels.

TESTBED AND INSTRUMENTATION - Ricardo routinely perform FTP transient certification and development testing on a purpose-built transient testbed (18). One of the main aims of the work described in this paper was to develop techniques for the optimisation of advanced electronically controlled engine management systems using existing testbeds. A conventional steady-state installation was therefore used as the starting point, with additional instrumentation added as and when required.

The engine installation incorporated a Schenck W400 eddy-current dynamometer. There was no facility for motoring and the four motoring points in the test cycle were approximated by no-load operation at the setpoint speeds. The first phase of the work was therefore to develop a speed and torque controller to enable the engine to be run over the NYNF[2] cycle.

The engine was instrumented to measure governor demand lever, rack and VGT guide vane positions, inlet and exhaust system pressures and temperatures and turbocharger speed. Engine torque was calculated from the measured brake torque and the load inertia torque.

Gaseous exhaust emissions were measured using instrumentation routinely used for steady-state testing (non-dispersive infra-red analysers for CO and CO_2, flame ionisation detector for total HC, chemiluminescent analyser for NOx and paramagnetic analyser for O_2).

Smoke was measured using a Celesco in-line smokemeter and particulates were measured using a tapered element oscillating microbalance (TEOM), see (19), sampling from a constant volume flow rate mini-dilution tunnel. Between 2% and 6% of the engine exhaust was sampled into the mini-dilution tunnel. The sampling system was characterised by measuring the dilution ratio over the engine speed and load range under steady-state conditions. Under transient conditions the instantaneous raw and dilute NOx were measured to verify the exhaust split. Tunnel temperatures were between $40^{\circ}C$ and $51^{\circ}C$ during the transient tests.

Transient compressor airflow was derived from measurements of the instantaneous pressure drop across a viscous flow air meter. The difference between the compressor flow and engine airflow during transients was calculated using the rate of change of inlet system pressure and temperature. As a cross-check transient engine air flow was also calculated from measurements of engine speed and boost pressure and temperature, using estimated volumetric efficiencies from steady-state measurements.

The transient fuel flow rate was derived from measurements of engine speed and rack position using the steady-state fuel pump map. Integration of this instantaneous fuel flow rate produced good agreement with the integrated fuel consumption measured gravimetrically.

Overall engine air/fuel ratio was calculated from the reconstructed exhaust CO_2 measurement. This was cross-checked against the airflow and fuel flow measurements to validate both the reconstruction procedure and the transient air/fuel ratio.

SPEED AND TORQUE CONTROLLER - The controller design procedure was first used in the development of a GPC speed and torque controller.

The testbed was considered as a two input (speed demand, brake torque demand), two output (engine speed, brake torque) system, see Figure 1. The GPC controller was designed to minimise a finite horizon quadratic cost function of the form:

$$J = [W(t) - \hat{X}(t)]^T Q [W(t) - \hat{X}(t)] + U^T(t) R U(t) \ldots\ldots(1)$$

where

$W(t)$ = future setpoints $W(t) = [w^T(t+1) \; w^T(t+2) \ldots w^T(t+n)]^T$
$\hat{X}(t)$ = future predicted states $\hat{X}(t) = [\hat{x}^T(t+1) \; \hat{x}^T(t+2) \ldots \hat{x}^T(t+n)]^T$
$U(t)$ = current and future control inputs $U(t) = [u^T(t) \; u^T(t+1) \ldots u^T(t+j-1)]^T$

J = cost function
Q = state weighting matrix
R = control weighting matrix
n = prediction horizon
j = control horizon
t = time

and superscript T denotes matrix transposition.

The controller minimised the current and future weighted errors between the setpoint speeds and torques, W, and the predicted speeds and torques, X. The identified models were used to predict the future performance of the system. The weighted controller actions, U, were also included in the cost function to regulate the activity of the controller. The controller was designed as a multivariable system with weighting, Q, chosen to achieve good speed and torque control and good decoupling between the speed and torque loops.

Due to the large magnitude of the speed and torque changes experienced during the NYNF[2] test section some compensation for the non-linearities of the system was required. The steady-state governor and dynamometer characteristics were therefore mapped to make the controlled system linear under steady-state conditions. This 'linear' system was used during the system identification process. Note that the speed and torque controller was designed to be 'robust' enough to handle large variations in engine performance. It therefore required no information about VGT GVP changes, see Figure 1.

Figure 2 compares the setpoint and measured speed, torque and power with a fixed VGT GVP set to give a turbine area equivalent to a Holset 4LGK turbocharger. Correlation procedures specified in the FTP regulations (20) were used to verify that the engine was running the cycle within acceptable limits.

The engine performance only deviated significantly from the setpoints during the initial load application. As the engine could not achieve the setpoint torques, the GPC controller was automatically switched to a second controller. This regulated the engine speed, via the brake torque, while the rack was held on the maximum (boost controlled) limit. The switching between controllers caused the disturbances in torque shown in Figure 2. With further refinement in gain switching a smoother changeover may have been achievable. The problem of rapid transient engine torque development appears with most highly rated turbocharged engines. The inability of the engine to achieve the setpoint torque is a function of the engine rather than the controller and commonly occurs on sophisticated testbeds with DC dynamometers.

A reduction in turbine area improved the boost and torque response. For this reason the integrated power over the cycle varied from test to test from 1.5% to 5.0% below that of the setpoint cycle. Specific emissions have therefore been compared.

ANALYSER RECONSTRUCTION - Analysers suitable for instantaneous measurements of exhaust emissions (CO, CO_2, HC, NOx) are not generally available. While fast response flame ionisation detectors (FID's) have recently come on the market (21) and oxygen sensors are widely used in gasoline applications, no equivalent NOx analysers currently exist.

Analyser systems used for steady state test work normally have heavily damped characteristics to ensure good stability. The physical configuration of most testbed layouts and sampling systems also introduce substantial transport delays. The identification and modelling techniques were therefore used to develop reconstruction filters for the analysers. These enabled the instantaneous exhaust emission concentrations to be calculated from the analyser output signals. Some measure of exhaust flow was required to calculate the mass related emissions. This was derived from airflow, engine speed and rack position measurements.

The reconstruction techniques were based on a finite horizon theory closely related to the GPC control theory described above. A detailed description is beyond the scope of this paper and will be given in a future publication.

VGT CONTROLLERS - Two types of VGT controller have been considered:

1. Steady-state map following controllers that determined VGT GVP as a function of engine speed and load. Two maps were used. The first (FE) map gave the optimum GVP for best fuel economy, see Figure 3a. The second (LE) map was designed to minimise NOx/HC/particulate emissions, at the expense of fuel economy if necessary, see Figure 3b.

2. A controller designed using advanced GPC techniques that calculated the transient VGT GVP for best NOx/particulate emissions based upon models of the engine system. This used feedback from parameters that could practically be measured in a commercial vehicle (ie. measuring governor demand lever position, rack position, engine speed, boost pressure, airflow, turbocharger speed and VGT GVP), see Figure 1.

The range of VGT modulation was limited for all the controllers by constraints of peak cylinder pressure, turbocharger speed, compressor surge/choke and turbine inlet temperature.

The GPC VGT controller was designed using the same techniques previously described for the speed and torque controller. A series of PRBS tests were performed to identify models describing the effect of speed demand, brake torque demand and turbine area demand on engine performance and emissions. Smoke and HC were used as measures of exhaust particulates during the VGT controller design stage.

Mathematical models were developed relating the smoke (at high loads) and the HC (at light

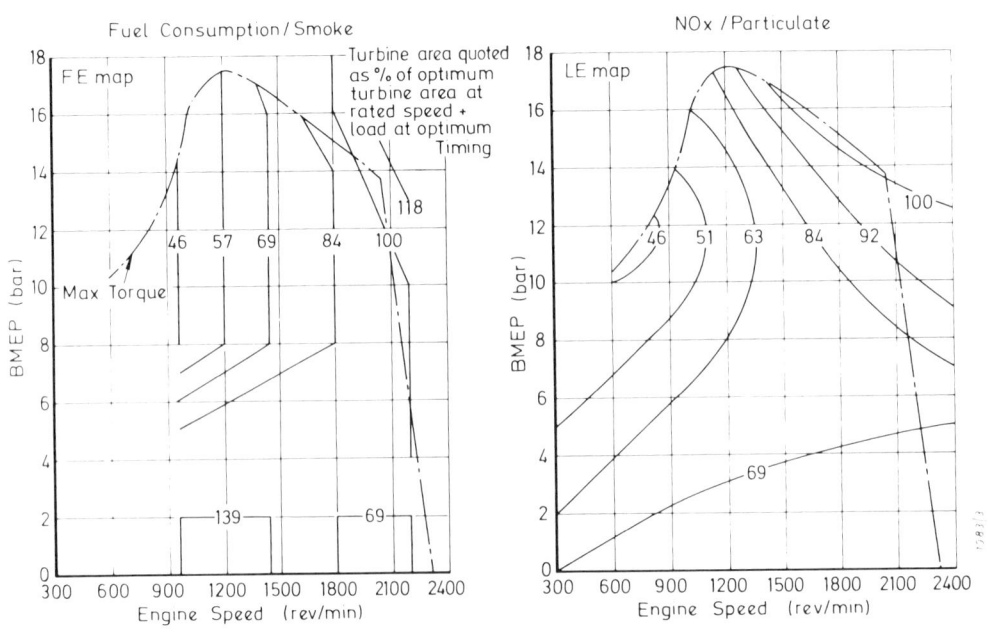

Figure 3. VGT Maps for Fuel Economy and Low Emissions.

loads) to measurable engine parameters, see Figure 1. Similar models were developed for the boost and exhaust pressures and turbocharger speed. Analysis of the models led to a boost maximising VGT controller which used the following model:

Controllable input : Turbine area demand.
Disturbances   : Governor demand lever position, rack position and engine speed.
Outputs   : Turbocharger speed, boost pressure, airflow.

In order to avoid the operational constraints of the VGT, the controller was designed to maximise boost pressure but operate at, or below, 90% of the maximum allowable turbocharger speed.

TEST RESULTS

BASELINE - A baseline transient test was first performed with the VGT GVP fixed to give a turbine effective area similar to that of the Holset 4LGK turbocharger.

Figure 4 shows the gaseous emissions (CO_2, HC, NOx) as measured by the analysers. The transport delays have been removed from the analyser signals and the reconstructed emissions are shown for comparison. The differences between measured and reconstructed HC are small compared to the distortion caused by the NOx analyser. Figure 4 also shows the instantaneous air/fuel ratio calculated from the reconstructed

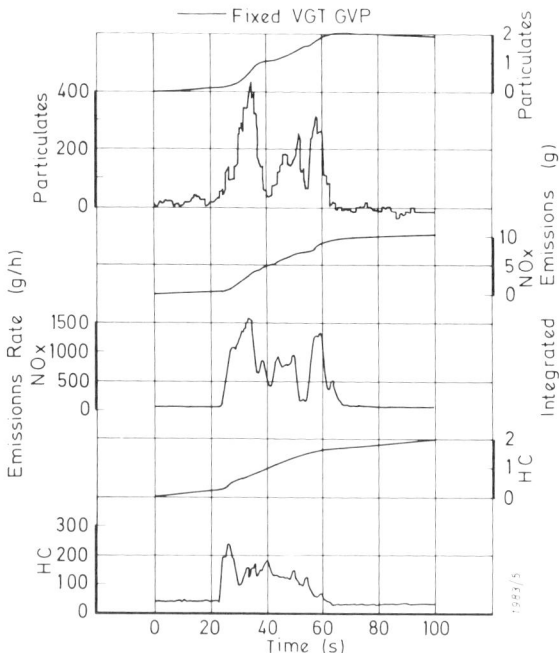

Figure 5. Reconstructed HC, NOx and Particulate Emissions over the NYNF[2] Cycle with Fixed VGT Guide Vane Position.

CO_2 which is compared with that derived using the measured airflow and fuel flow rate calculated from the engine speed and rack position. The agreement is very good over a wide range of air/fuel ratios.

Reconstructed HC and NOx emissions and exhaust flow rate were used to calculate the mass emissions shown in Figure 5. Also shown are particulate emissions derived from TEOM measurements using the reconstructed dilution ratio. The instantaneous NOx shows similar trends to the torque signal, Figure 2. Figure 6 shows the transient boost pressure, rack position, turbocharger speed, VGT GVP and smoke opacity. Smoke was regulated by the standard boost controlled rack limiter. Comparison of the smoke and HC with particulates from TEOM measurements indicates that the first particulate peak comprises of HC and carbon, the intermediate section is only HC and the final peak is mostly carbon.

Analysis of the TEOM particulate filters for the baseline tests using gas chromatography (22) indicated that 26% of the particulate was hydrocarbon based. Approximately half of this was unburned oil and half unburned fuel.

MAP FOLLOWING VGT CONTROLLERS - Figures 6 and 7 show results for the transient test with the VGT modulated to follow the steady-state FE map shown in Figure 3a. The results are superimposed on the data from the fixed GVP tests for comparison purposes. It can be seen that the turbine area was increased at idle to reduce pumping work and was modulated over a large operating range during the test cycle.

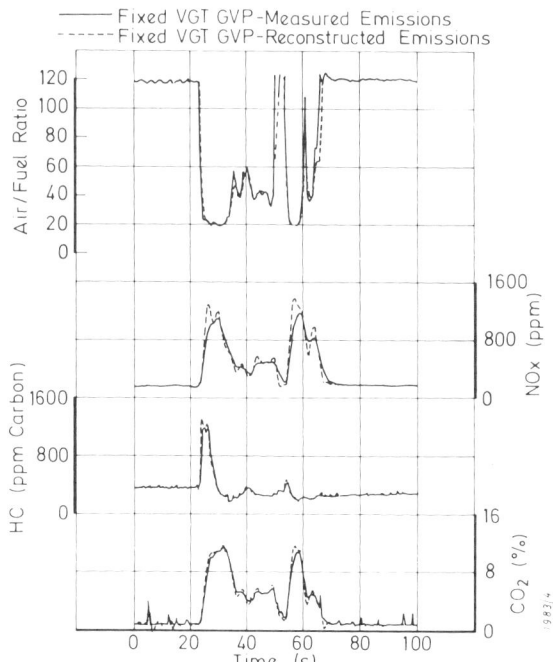

Figure 4. Measured and Reconstructed Gaseous Emissions and Air/Fuel Ratio over the NYNF[2] Cycle with Fixed VGT Guide Vane Position.

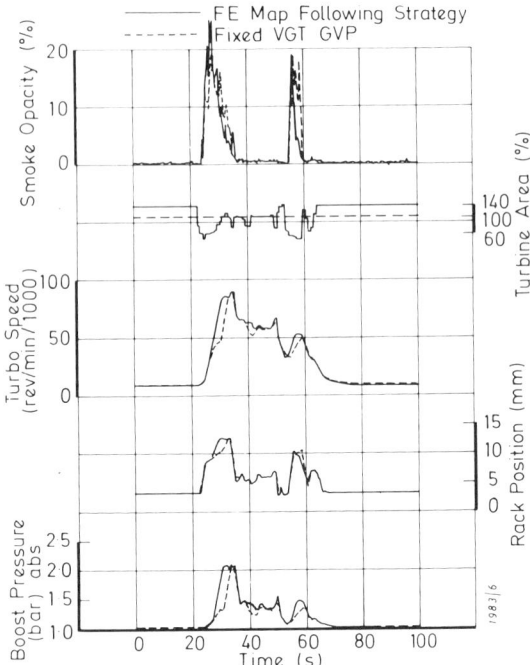

Figure 6. Comparison of Boost Pressure, Rack Position, Turbocharger Speed, VGT GVP and Smoke with Fixed and Fuel Economy Guide Vane Position Strategies.

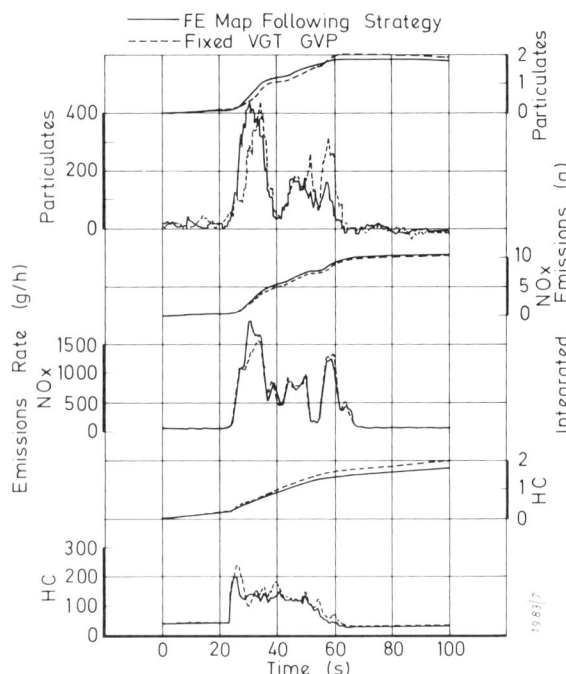

Figure 7. Comparison of HC, NOx and Particulate Emissions with Fixed and Fuel Economy VGT Guide Vane Position Strategies.

Figure 6 shows that the boost pressure and turbocharger speed were significantly increased during the later stages of the load applications, but the smoke peaks were not reduced, ie. the turbocharger response was too slow to increase the minimum air/fuel ratio. Higher boost pressure later in the test reduced smoke significantly.

Figure 7 compares the HC, NOx and particulate mass emissions with the baseline test. The smoke and particulates were worse during the initial load application with the FE map but HC's were improved somewhat over most of the test. The main reduction in particulates was due to reduced smoke during the later parts of the cycle. Smoke and NOx were higher during the initial load application because the turbocharger developed more boost, and the engine more torque, with the rack limiter operational. During this phase, the speed and torque controller caused the rack to limit because the engine torque was below the setpoint. Overall there were reductions of 15% in specific HC and 23% in average smoke opacity with the FE strategy. Specific particulates were reduced by 12%. Fuel economy was marginally better and airflow was increased by 4%.

Figures 8 and 9 show similar diagrams for the LE map following strategy shown in Figure 3b. Figure 8 shows that the turbine area was reduced at light loads to reduce HC emissions. This resulted in almost constant turbocharger acceleration and increased torque with the rack limiter operational. However at the idle condition the exhaust energy was low, even with reduced turbine area, and higher engine back pressure penalises engine performance. Note that the initial build up in boost pressure is not as great as with the FE strategy, Figure 6, and that the turbine area could have been reduced for longer without exceeding the speed limit.

Figure 9 shows the effect on HC, NOx and particulate mass emissions. Reduction in particulates was partly due to reduced smoke during the later stages of the two main load applications. HC was significantly reduced over the entire transient portion of the test, but particularly during the initial load application, ie. as the fuelling was rapidly increased from idle. Overall specific HC was reduced by 20% and average smoke opacity by 29% with the LE strategy. Specific particulates were reduced by 13%. Fuel economy was marginally worse but airflow was increased by 10%

GPC CONTROLLER - The GPC controller was designed to increase the transient air/fuel ratio to reduce smoke during the high load parts of the cycle and reduce HC emissions through reduced ignition delay and improved transient fuel/air mixing.

Preliminary tests showed that the controller reduced turbine area to the minimum allowed over most of the test, only opening it up to avoid turbocharger overspeed. This resulted in a very rapid initial turbocharger acceleration and boost pressure rise from the initial idle condition. However this increase in boost pressure was not accompanied by a

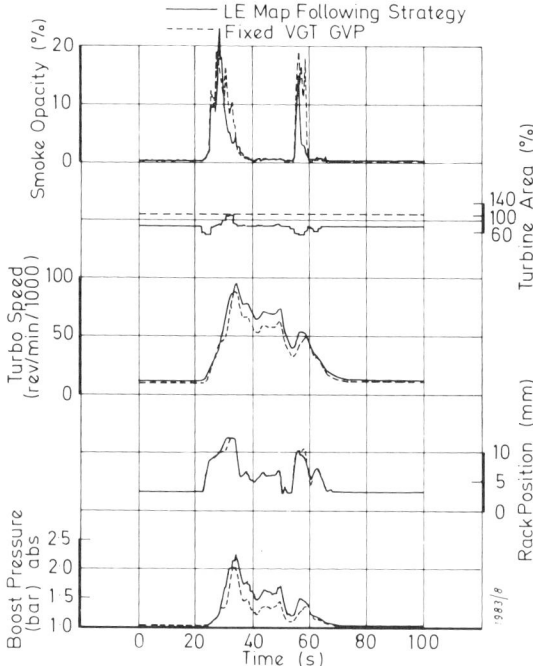

Figure 8. Comparison of Boost Pressure, Rack Position, Turbocharger Speed, VGT GVP and Smoke with Fixed and Low Emissions VGT Guide Vane Position Strategies.

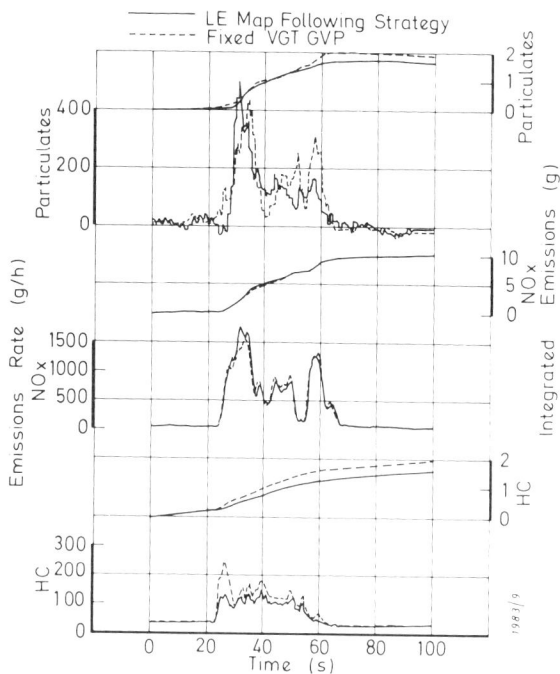

Figure 9. Comparison of HC, NOx and Particulate Emissions with Fixed and Low Emissions VGT Guide Vane Position Strategies.

proportional increase in airflow, due to the excessive engine back pressure created by the shut-down turbine. The result was that the standard (boost controlled) rack limiter made the air/fuel ratio too rich and significantly increased the initial smoke. Later in the cycle the smoke was reduced.

The rack limiter was adjusted to reduce the allowable rack travel for a given boost pressure. The VGT would therefore have to increase boost pressure to maintain the baseline torque curve. Tests showed that this significantly improved the transient smoke and that the engine performance was closer to the setpoint cycle than any of the other configurations tested. In contrast, the changes to the rack limiter also improved the smoke for the baseline fixed GVP build, but the loss of transient torque due to the poor turbocharger response meant that the average engine power was reduced and the cycle failed to meet the FTP validation criteria. Some further improvement in the smoke using the map-following strategies might have been achieved, but these maps are largely a compromise and finding alternatives that improve transient boost response yet work safely under steady-state conditions would have been a time consuming exercise.

Figures 10 and 11 show test results using the GPC VGT controller. The effect of the GPC control strategy on boost pressure, rack position, turbocharger speed and smoke is again compared with the baseline engine performance. The higher initial turbocharger speed and almost constant acceleration during the first part of the test, resulted in improved boost pressure response and increased engine torque compared to the baseline test. The GPC controller maintained the turbocharger speed close to 90000 rev/min over the considerable portion of the cycle where sufficient turbine energy was available, despite very large engine speed and load changes. Specific air consumption was increased by 18%.

Increased rack travel (fuelling) during idle and the mid part of the test clearly shows the fuel economy penalty for increased boost pressure. Specific fuel consumption was increased by 6% compared to the baseline engine. A reduction in boost would have improved fuel consumption during the mid section of the test, but worsened HC emissions. The boost level must also be maintained high enough to limit the smoke during the second rapid load change.

Figure 11 shows that HC emissions were reduced by 37% over the test. Only 5% of this reduction was due to lower steady-state HC at idle. The reduced smoke (34%) and HC resulted in a 34% reduction in particulates compared to the baseline test. As with the case of the map-following VGT controllers there was no significant change in the NOx emissions over this cycle.

Table 1 compares the overall performance of the baseline, map following and GPC controllers over the NYNF[2] test.

87

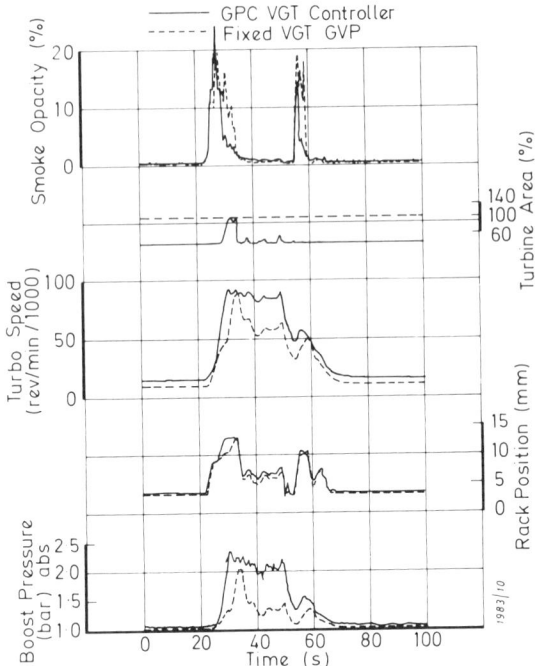

Figure 10. Comparison of Boost Pressure, Rack Position, Turbocharger Speed, VGT GVP and Smoke with Fixed and GPC VGT Guide Vane Position Strategies.

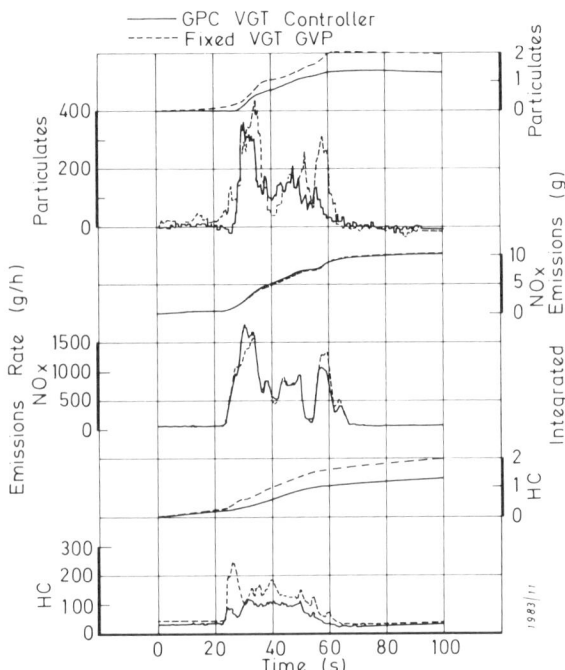

Figure 11. Comparison of HC, NOx and Particulate Emissions with Fixed and GPC VGT Guide Vane Position Strategies.

Table 1 Comparison of Fuel Consumption and Emissions with Fixed Turbocharger GVP, Steady-State Map following and GPC VGT Strategies

GVP Strategy	Fixed	FE	LE	GPC
Average Power (kW)	56.65	55.41	56.16	57.49
Fuel Consumption (g/bhp.h)	191.2	190.5	191.4	202.6
Air Consumption (kg/bhp.h)	7.44	7.70	8.14	8.75
Average Smoke (%)	2.47	1.90	1.77	1.64
HC (g/bhp.h)	1.15	0.98	0.92	0.72
NOx (g/bhp.h)	6.52	6.43	6.48	6.16
Particulates (g/bhp.h)	1.32	1.17	1.14	0.87

Particulate emissions over the NYNF[2] section account for 15-20% of the total FTP cycle particulate emissions. A 34% reduction in particulates over this section alone would translate into a 4-6% reduction over the full cycle.

DISCUSSION

This paper has outlined a systematic approach to the development and optimisation of controllers for testbed and vehicle applications. Significant reductions in smoke, HC and particulate emissions have been achieved through the development of a GPC controller for a VGT. The results also indicate that the VGT control must be integrated with fuelling control to achieve the maximum benefits.

The work has been performed on an engine that has emissions levels corresponding to a US 1990 specification engine. A 1991 engine would have a higher compression ratio, better oil control and lower smoke levels and consequently the benefits of VGT alone for emissions control would not be as great.

As electronically controlled FIE with fuelling, timing and rate control become more common, integrated control strategies will be required for both steady-state and transient operation. Some form of systematic transient optimisation will be an essential part of any 1994 development programme. The increased flexibility and complication will demand a systematic approach to future controller designs.

As part of an on-going research programme Ricardo will apply the methodology described in this paper to a 1991 specification engine incorporating electronically controlled fuelling and injection timing in addition to VGT. The objective will be to optimise both exhaust emissions and fuel economy. The results of this programme will indicate the potential reductions in 'engine-out' emissions levels and the exhaust aftertreatment requirements to achieve US 1994 limits.

CONCLUSIONS

This work has demonstrated the influence of transient VGT control strategies on fuel consumption and emissions for a heavy-duty diesel engine.

A systematic approach to the controller design procedure has been used to develop both a testbed speed and torque controller and a VGT controller suitable for vehicle applications.

A 'first-generation' VGT controller has been designed using advanced control techniques. The objective for this controller was to minimise particulate emissions without increasing NOx.

Engine tests were carried out over a highly transient section of the FTP test cycle with a high particulate contribution. Compared to the baseline build, with the VGT set to give a turbine area similar to that of the standard fixed geometry machine, the following results were achieved:

1. Reductions in HC, smoke and particulate emissions of up to 20%, 29% and 13% respectively using map following VGT strategies.

2. Reductions in HC, smoke and particulate emissions of up to 37%, 34% and 34% respectively using a GPC VGT controller designed using advanced control techniques. This alone would translate into a 4-6% reduction in particulates over the full FTP cycle.

3. The GPC controller increased specific airflow by 18% and resulted in a deterioration in fuel economy of 6%.

The use of advanced control techniques has been shown to offer a cost-effective and practical approach towards the design of integrated multivariable controllers for future powertrain applications.

ACKNOWLEDGEMENTS

The authors would like to thank the Directors of Ricardo Consulting Engineers Ltd for permission to publish this paper. Sincere thanks are due to a large number of colleagues whose contributions are reflected in the contents of the paper.

REFERENCES

1. FLAXINGTON D, SZCZUPAK D T
"Variable Area Radial-Inflow Turbine"
I Mech E Conference: Turbocharging and Turbochargers, C36/82, 1982.

2. OKAZAKI Y, MATSUDAIRA N, HISHIKAWA A
"A Case of Variable Geometry Turbocharger Development"
I Mech E Conference: Turbocharging and Turbochargers, C111/86, 1986.

3. ROESSLER M, SWENSON K R
"Variable Nozzle Turbochargers for Medium-Speed Engines"
SAE 880119, 1988.

4. SATOH H, MIYAUCHI J, NAKAZAWA N, & MATSUO E
"Development of a Variable Geometry Turbocharger for Trucks and Buses"
GTSJ 1983 Tokyo International Gas Turbine Congress, 83-TOKYO-IGTC-77, 1983.

5. ARNOLD S
"Schwitzer Variable Geometry Turbo and Microprocessor Control Design and Evaluation"
SAE 870296, 1987.

6. NORBYE J P
"How Future Turbochargers Take Shape at KKK"
High Speed Diesel Report, Nov-Dec 1988.

7. NEEDHAM J R, SANDFORD M H
"A Truck Engine for the 1990's"
ASME 87-ICE-50, 1987.

8. WATSON N, BANISOLEIMAN K
"Performance of a Highly Rated Vehicle Diesel Engine with a Variable Geometry Turbocharger"
I Mech E Conference: Turbocharging and Turbochargers, C103/86, 1986.

9. FRANKLIN P C, WALSHAM B E
"Variable Geometry Turbochargers in the Field"
I Mech E Conference: Turbocharging and Turbochargers, C121/86, 1986.

10. HASHIMOTO T, OKADA K, OIKAWA T
"ISUZU New 9.8L Diesel Engine with Variable Geometry Turbocharger"
SAE 860460, 1986.

11. McCUTCHEON A R S, BROWN M W G
"Evaluation of a Variable Geometry Turbocharger Turbine on a Commercial Diesel Engine"
I Mech E Conference: Turbocharging and Turbochargers, C104/86, 1986.

12. WALLACE F J, HOWARD D, ROBERTS E W, ANDERSON U
"Variable Geometry Turbocharging of a Large Truck Diesel Engine"
SAE 860452, 1986.

13. WATSON N, BANISOLEIMAN K
"A Variable-Geometry Turbocharger Control System for High Output Diesel Engines"
SAE 880118, 1988.

14. WINTERBONE D E, JAI IN S
"Control Studies of an Automotive Turbocharged Diesel Engine with Variable Geometry Turbine"
SAE 880485, 1988.

15. NOBLE A D, BEAUMONT A J, MERCER A S
"Predictive Control Applied to Transient Engine Testbeds"
SAE 880487, 1988.

16. McCLURE B T
    "Characterization of the Transient Response of a Diesel Exhaust-Gas Measurement System"
    SAE 881320, 1988.

17. CLARKE D W, MOHTADI C, TUFFS P S
    "Generalised Predictive Control,
    Part 1: The Basic Algorithm.
    Part 2: Extensions and Interpretations"
    Automatica, Vol. 23, No. 2, pp 137-160, 1987.

18. HALES J M, MAY M P
    "Transient Cycle Emissions Reduction at Ricardo - 1988 and Beyond"
    SAE 860456, 1986.

19. SHORE P R, CUTHBERTSON R D
    "Application of a Tapered Element Oscillating Microbalance to Continuous Diesel Particulate Measurement"
    SAE 850405, 1985.

20. EPA Code of Federal Regulations, 40 CFR, Parts 81-99, July, 1987.
    Section 86.1341-84 Test Cycle Validation Criteria.

21. COLLINGS N, EADE D
    "An Improved Technique for Measuring Cyclic Variations in the Hydrocarbon Concentration in an Engine Exhaust"
    SAE 880316, 1988.

22. CUTHBERTSON R D, SHORE P R, SUNDSTROM L, HEDEN P-O
    "Direct Analysis of Diesel Particulate-Bound Hydrocarbons by Gas Chromatography with Solid Sample Injection"
    SAE 870626, 1987.

APPENDIX A   Engine Specification

Make and Model	: Volvo TD120F
Type	: 4 stroke, turbocharged, aftercooled, diesel
Combustion system	: Swirling, direct injection, toroidal bowl
Bore and stroke	: 130.2 mm by 150.0 mm
Configuration	: 6 cylinder in-line
Swept Volume	: 12 litres
Compression ratio	: 14.2 (nominal)
Injection Timing	: $15°$ CA BTDC (spill)
Maximum power	: 285 kW at 2050 rev/min (13.9 bar BMEP)
Maximum torque	: 1660 Nm at 1200 rev/min (17.4 bar BMEP)
Idle speed	: 600 rev/min
Fuel pump	: Bosch P7100
Governor	: Bosch RQV combination type
Injectors	: Bosch 5 x 0.34 mm, $150°$ cone angle
Turbocharger	: IHI RHC9V VGT, TCW58 compressor, 1700V single-entry turbine

890396

An Investigation of Cylinder Pressure as Feedback for Control of Internal Combustion Engines

Peter Wibberley and Christopher A. Clark
Ricardo Consulting Engineers Ltd.

ABSTRACT

The advantages of closed loop over open loop control systems are generally recognised. However, existing engine management systems implement most control functions in open loop because suitable feedback sensors are not available. Even for so-called closed loop air fuel ratio controllers, shortcomings of the exhaust gas oxygen (EGO) sensor limit the potential effectiveness of closed loop control. A more direct measure of the combustion process, such as cylinder pressure, can yield sufficient information for the closed loop operation of many of the combustion control functions; this paper presents the results of a prediction algorithm which can derive a variety of feedback signals from cylinder pressure.

Cylinder pressure, together with several combustion variables, including air-fuel ratio, exhaust gas recirculation rate, and NO_x, HC, CO and CO_2 emissions were measured at various operating points. The combustion variables were assumed to be a function of cylinder pressure and the parameters of this function were identified from the experimental data. Predictions of combustion variables were generated from individual cycles of cylinder pressure data using this function. These predictions show good correlation with the measured values and could form the basis of a full authority, high performance closed loop combustion controller. Potential benefits include reduced exhaust gas emissions, improved fuel economy and reduced roughness.

The technique can also be applied to data from other combustion sensors and combustion parameters. Thus other combustion sensors and additional combustion control functions could be considered for a practical engine management system.

*Numbers in parentheses designate references at end of paper.

MOST CURRENT ELECTRONIC ENGINE management systems, including those described as being closed loop, employ fuelling and ignition strategies which are essentially open loop. Maps of fuelling or ignition settings are indexed, usually by engine speed and by some other variable strongly related to engine load, e.g. manifold air pressure. Compensation may also be applied for coolant temperature and ambient pressure, and during engine transients further correction factors are required (1,2)*.

The values tabulated in the maps provide some form of compromise between various objectives, such as emissions and fuel economy. Setting up the map involves running an engine on a testbed or chassis dynamometer across a wide range of operating conditions; fuelling and ignition settings are chosen at each operating point in accordance with the required compromise.

Successful though this approach has been, there are a number of drawbacks:

(1) The mapping process is tedious and error prone; transient correction factors can particularly difficult to calibrate;

(2) The setting chosen at each operating point may not be optimal. The values may be chosen using rules of thumb rather than by accurately evaluating a trade-off between the various objectives;

(3) The settings chosen may be a good compromise for the whole engine, but non-ideal for each individual cylinder. Significant variations exist between the characteristics of the different cylinders of an engine and improvements would result from controlling each cylinder separately;

(4) Ideal maps for the tested engine may be non-ideal for production engines, due to variations between individual engines of the same design;

(5) An ideal map for a given engine will become less appropriate as the engine components wear, and its characteristics change.

Closed loop or feedback control systems can overcome some of these problems. The output of the controller is based, not merely on predetermined values, but on a comparison of the actual and desired responses. Thus discrepancies between the two responses can be corrected and the effect of variations in the engine are reduced.

The operation of any closed loop controller depends on having a measure of the variables to be controlled, or of variables closely related to them. On gasoline engines fitted with a three way catalyst it is common practice to use an exhaust gas oxygen (EGO) sensor; the sensor gives an indication of whether the engine is running at or near stochiometric. This form of closed loop engine management system, however, still has significant shortcomings:

(a) EGO sensors give limited information concerning the combustion process and hence can only be used to control limited aspects of it. For instance, EGR rate control is not possible;

(b) There is a time delay before changes in in-cylinder air fuel ratio, i.e. a required measurement, can be detected by the EGO sensor;

(c) The sensor itself generally has a poor transient response;

(d) Cylinder specific information is largely lost due to the location of the sensor and the mixing of exhausts which takes place downstream of the manifold.

Thus current closed loop engine management systems can enable fuelling corrections should the engine run consistently lean or consistently rich, and the effects of aging and of the variation between engines (in regard of fuelling) can be compensated. However other facets of combustion control are still open loop; and even for fuelling, cylinder to cylinder variations, and deviations between actual and desired air fuel ratios occurring during engine transients, cannot be corrected.

So although the advantages of closed loop control are recognised, there are practical limitations to the effectiveness of its application in current engine management systems. Whilst research and development of a wide range of sensors is taking place (3), a more direct measure of the combustion process, such as cylinder pressure, could yield far more feedback information than existing systems, and also avoid some of the inherent problems. With feedback information derived from cylinder pressure, other control functions (EGR rate, boost pressure, etc.) could also be closed loop controlled.

This paper describes the results of a prediction algorithm initially designed to extract air fuel ratio and EGR rate feedback signals from a cylinder pressure diagram. The technique is extended to predict other combustion variables, namely exhaust emissions.

THEORY

The cylinder pressure diagram, sampled at degree intervals throughout an engine cycle, can be represented as a vector,

$$\underline{x} = [x_1\ x_2\ x_3\ \ldots\ x_{720}]$$

Air fuel ratio, EGR rate, ignition timing and NO_x, HC, CO and CO_2 emissions can be represented as a vector of combustion parameters,

$$\underline{y} = [y_1\ y_2\ y_3\ \ldots\ y_n]$$

We assume that values of $\underline{x}$ and $\underline{y}$ measured simultaneously are functionally related. Previous work (4) used semi-heuristic arguments in first taking various moments of the cylinder pressure diagram, and then identifying a relationship between these and air fuel ratio and EGR rate. In this study, the combustion parameters are expressed as a function of the entire cylinder pressure diagram, avoiding any information loss in the process of taking moments. Hence we assume only,

$$\underline{y} = f(\underline{x})$$

Of course cycle by cycle measurements of $\underline{y}$ are not usually available, even on the testbed. Hence the measured mean value for an operating point is taken as being representative of each individual cycle measured at that operating point.

Calculating the prediction algorithm involves identifying a function $f(\underline{x})$ which adequately fits the experimental data. Validation of the predictor involves comparing predictions of $\underline{y}$ from measured cylinder pressure data with measured values of $\underline{y}$. Note that a separate data set must be used for this validation avoid falsely optimistic results.

ENGINE TESTS

Engine tests were performed using a Ricardo Hydra Mk. III, single-cylinder, fuel-injected gasoline engine with a swept volume of 0.496 litre, a compression ratio of 8.9:1 and a bathtub combustion chamber in the cylinder head.

Cylinder pressure was measured using a Kistler 6121 piezo-electric transducer, cleaned and recalibrated every 20 running hours, mounted in the cylinder head. The cylinder pressure signal was sampled every one degree crank angle using a Ricardo High Speed Data Acquisition unit. Emissions were measured using Analytical Developments NDIR analysers (CO, CO_2 and NO_x) and an IPM RS5 heated flame ionisation detector (HC).

Air-fuel ratios values were derived from the measured exhaust gas composition using the Spindt equation (5,6). Exhaust gas recirculation rates were calculated from the measurements of CO_2 concentration in the exhaust gas, air and the air/recirculated exhaust gas mixture.

The tests were conducted at an engine speed of 40 rev/s and 2.5 bar BMEP. One mixture loop with no EGR rate and EGR loops at stochiometric and lean air-fuel ratios were taken; the test matrix is shown in Figure 1. At each test point, 350 consecutive cycles of cylinder pressure were recorded, making a total of 5250 cycles for the 15 test points.

METHOD

The 350 cycles of cylinder pressure data from each test condition were converted into a format suitable for the MATLAB analysis package (7) running on a Sun workstation.

Each cycle was offset, to remove charge amplifier drift, by setting the pressure at non-firing BDC to zero. The mean of all the cylinder pressure diagrams recorded was calculated. This was then subtracted from each individual cycle, since, with the technique being used, it is the difference between cylinder pressure diagrams which is important in discriminating between the different values of combustion parameters at the various operating conditions. Examples of cylinder pressure diagrams, unmodified, and 'centred' (i.e. with the mean diagram removed) are shown in Figures 2 and 3.

Only the cylinder pressures recorded between inlet valve closing (IVC, 56 degrees ABDC) and exhaust valve opening (EVO, 56 degrees BBDC) were used, on the assumption that only these contain useful information regarding the combustion process. Each test point data set was then divided into two by separating 'even' cycles from 'odd' cycles, giving independent 'design' and 'test' sets to enable cross-validation.

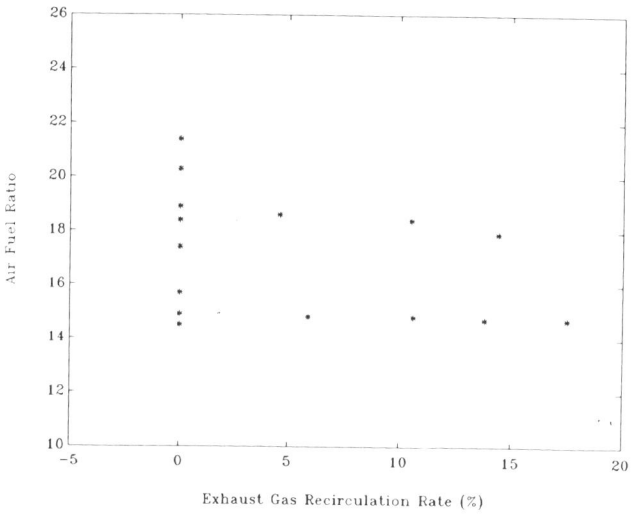

Figure 1 - Matrix of Engine Test Points

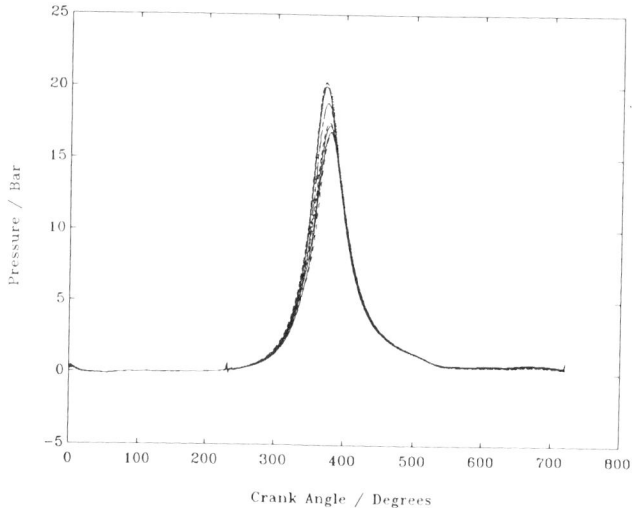

Figure 2 - Examples of Unmodified Cylinder Pressure Diagrams

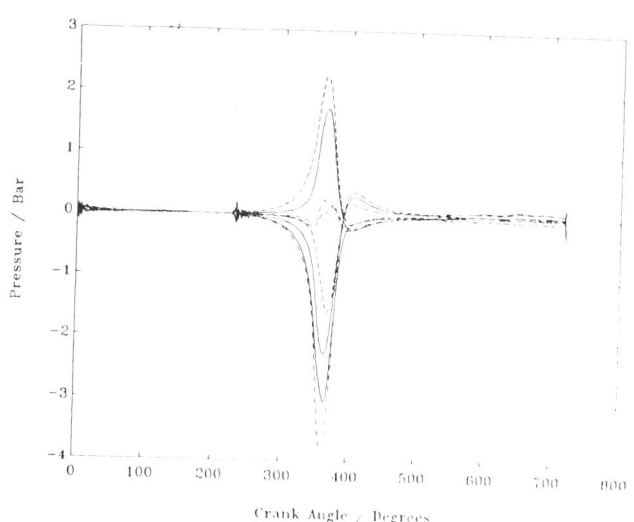

Figure 3 - Examples of Centred Cylinder Pressure Diagrams

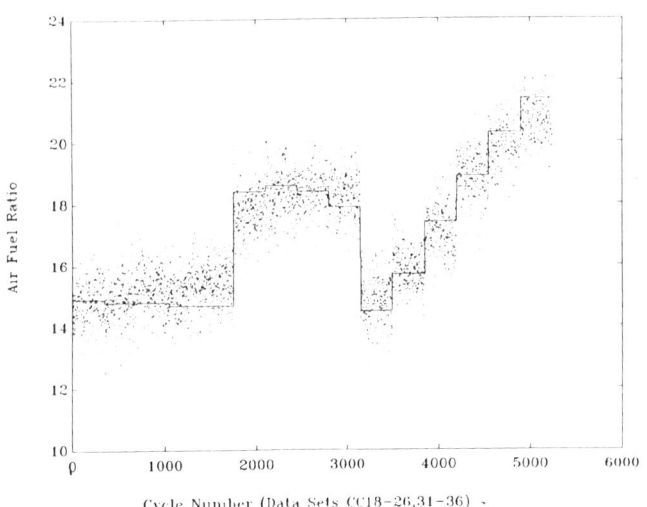

Figure 4 – Measured and Cycle by Cycle Predictions of Air Fuel Ratio (AFR)

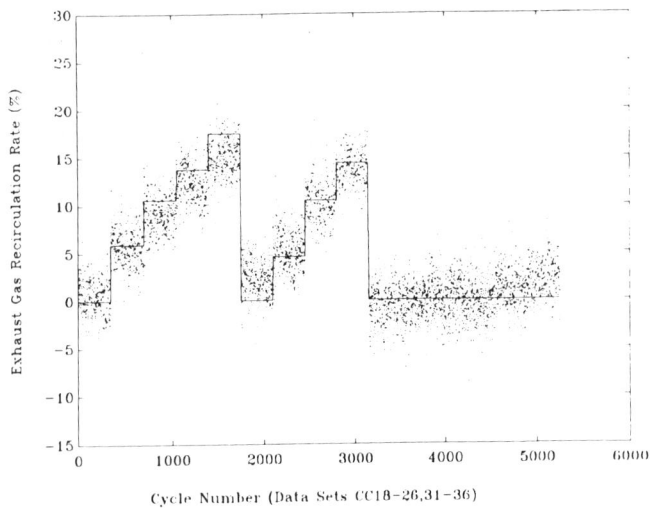

Figure 5 – Measured and Cycle by Cycle Predictions of Exhaust Gas Recirculation Rate (EGR)

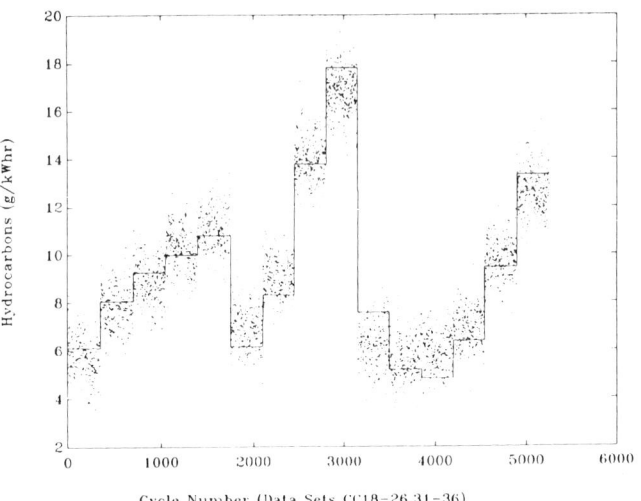

Figure 6 – Measured and Cycle by Cycle Predictions of Hydrocarbons (HC)

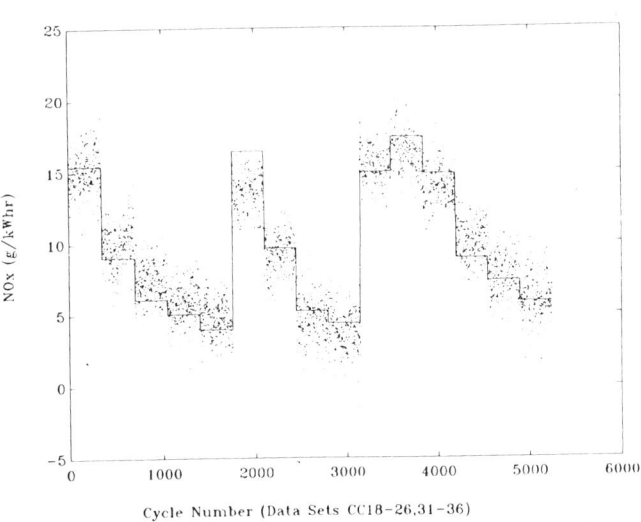

Figure 7 – Measured and Cycle by Cycle Predictions of NOx

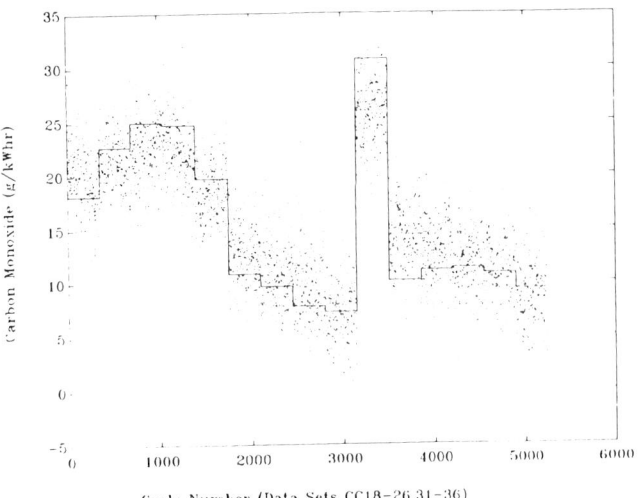

Figure 8 – Measured and Cycle by Cycle Predictions of Carbon Monoxide (CO)

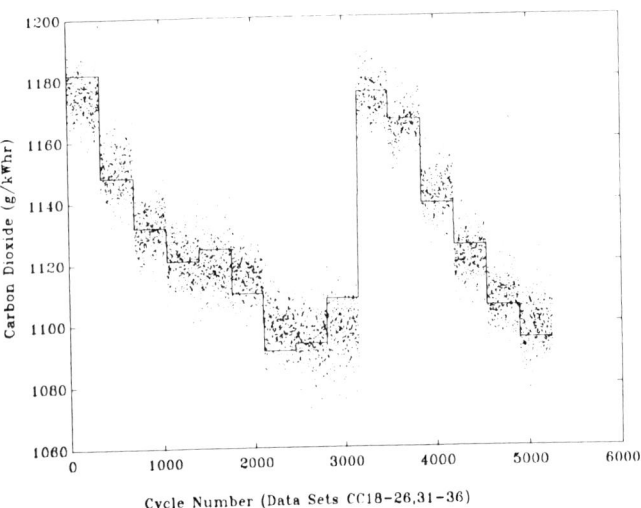

Figure 9 – Measured and Cycle by Cycle Predictions of Carbon Dioxide (CO_2)

RESULTS

The results are shown in Figures 4 to 9. These graphs show comparisons of measured and predicted values for each of the combustion parameters considered: AFR, EGR rate, HC, CO, CO_2 and NO_x.

Based on a single cycles of cylinder pressure data, the results comprise:

(i) Predictions based on a function $f(\underline{x})$ calculated from the odd cycles, cross-validated using the even cycles;

(ii) Predictions based on a function $f(\underline{x})$ calculated using the even cycles, cross-validated using the odd cycles.

In this way independence of the design set and the test set is maintained, but all the data is subjected to the validation test.

DISCUSSION

The predictions, though 'noisy' show excellent correlation with the measured values. The prediction errors can be ascribed to 3 sources:

(i) Cycle by cycle variations in cylinder pressure occurring despite constant operating conditions (AFR, EGR rate etc.);

(ii) Actual cycle by cycle variations in operating conditions (AFR, EGR rate etc.);

(iii) Errors in predicting combustion parameters due to shortcomings in the prediction process itself.

Whatever the error sources, the combustion parameter predictions can be regarded as being analogous to the output from a noisy sensor, in that the errors are substantially random and probably uncontrollable. Any control system using these predictions would be designed to reject such random variations.

The predictions are subject to some bias, i.e. the mean of the predictions at a given test point does not always agree accurately with the measured mean value at that test point. In any system seeking to control AFR or EGR rate, say, accurately to a target value this would present a problem. However, a conventional, slow response exhaust gas sensor could be used to correct for some prediction biases while the predictions would provide rapid feedback of transient changes.

Alternatively the emissions predictions could be employed in a controller designed to regulate emissions directly, rather than by maintaining target AFR and EGR rate. Here the absolute accuracy of the predictions is less important. This is because the objective of the controller would not be to achieve precise target values, but merely to minimise a combination of values (Figure 10).

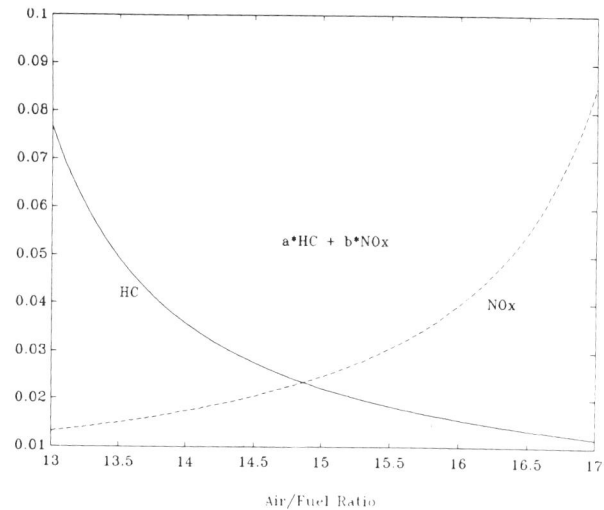

Figure 10 - AFR Required to Minimise a Function of Emissions is Dependent on Relative Trends and not Absolute Values

IMPLEMENTATION

The prediction algorithms are efficient, although the rate of cylinder pressure data acquisition is necessarily high. Predictions of several combustion parameters for all cylinders of an automotive gasoline or diesel engine would be well within the scope of current digital signal processing (DSP) processors; the same processor could also accommodate the closed loop control algorithms.

A schematic diagram of a closed loop controller employing combustion parameter predictions is shown in Figure 11. However, before a practical demonstrator of this form is developed, the robustness of the predictions under a variety operating conditions and against variations in engine build should be investigated.

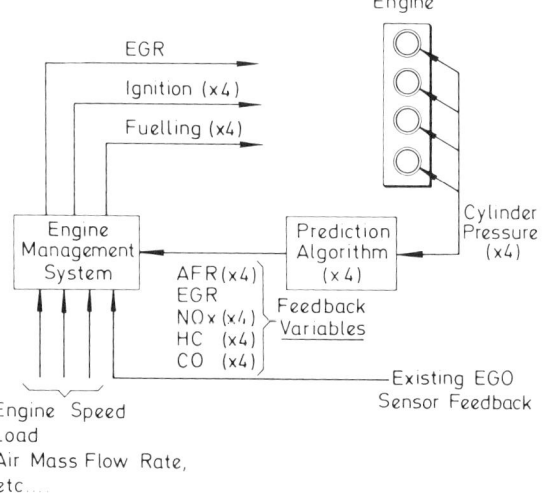

Figure 11 - Schematic Diagram of a Possible Implementation of the Prediction Algorithm within an Engine Management System

FURTHER WORK

Importantly, the data fitting techniques used are not specific to cylinder pressure nor to the combustion parameters used in this study. Any combustion sensor producing data correlating with the required parameters may be suitable; possible candidates might include optical combustion sensors, ionisation probes or even cylinder block accelerometers. Hence, in a production system, factors such as the sensor cost or robustness could be considered.

Predictors could also be designed for additional combustion parameters such as brake specific fuel consumption (BSFC) and controllers designed to optimise any one or combinations of these. Any combination of combustion sensor and combustion parameters can be explored by the analysis of data from engine tests where the appropriate data is gathered simultaneously.

In progressing towards a practical engine management system using these prediction techniques, first it will be necessary to repeat the results described across a wider range of engine operating conditions and on multi-cylinder engines. Secondly, subject to equally satisfactory results, a simulation of engine, predictors and control system would be evaluated. This would prove the potential benefits of the system, prior to its implementation in a demonstrator engine management system.

CONCLUSIONS

The use of a function fitting technique to calculate a predictor for various combustion parameters has been demonstrated. The correlation of the predictions with measured values is excellent for all the parameters investigated.

Although cylinder pressure is a very good source of information about the combustion process, the technique is not specific to this signal or the parameters investigated. Other sensors signals and parameters could be considered, with the strength of their correlation determining the likely degree of success.

The predictor eliminates the problems associated with poor sensor or measurement dynamics and could also provide feedback for parameters which cannot be measured using production quality sensors. The algorithms involved are efficient and the predictions could be used as feedback to enable tight closed loop control of combustion during transient and steady state operation and to optimise for emissions, fuel economy and roughness.

ACKNOWLEDGEMENTS

The authors would like to thank the Directors of Ricardo Consulting Engineers for permission to publish this paper. Sincere thanks are due to a large number of colleagues, especially those in the Ricardo Instrumentation and Control Laboratory, whose contributions are reflected in the contents of this paper. Particular thanks are due to Parag Vyas and Simon Beauchamp for performing much of the analysis work.

REFERENCES

(1) C.F. Aquino, "Transient A/F Control Characteristics of the 5 Liter Central Fuel Injection Engine", SAE 810494.

(2) S.D. Hires, M.T. Overington, "Transient Mixture Strength Excursions - An Investigation of Their Causes and the Development of a Constant Mixture Strength Fuelling Strategy", SAE 810495

(3) B.J. Challen, "Some Diesel Engine Sensors", SAE 871628.

(4) J.C. Gilkey, "Fuel-air Ratio Estimation From Cylinder Pressure in an Internal Combustion Engine", Ph.D dissertation (1984), Stanford University.

(5) R.S. Spindt, "Air-Fuel Ratio from Exhaust Gas Analysis", SAE 650507.

(6) L. Eltinge, "Fuel-Air Ratio and Distribution from Exhaust Gas Composition", SAE 680114

(7) "PRO-MATLAB User's Guide", The MathWorks Inc. (1987).

890397

A Complete Engine Diagnostic System for Military Application

C. Operti
IVECO
W. Duss and J.W. Freestone
DERECO

ABSTRACT

The possibility of fast identification of defective components and the reduction of repair time of vehicles is a common requirement for military applications.

In the development of a new power-pack for a tank application IVECO has included a complete diagnostic system that provides the following performances:

- at first level (i.e. on the field) the system provides indications on simple repair operations or the indication of a seriuos power-pack fault, such to require power-pack replacement. This level is fully supported on the tank itself and requires no external support.

- at second level (i.e. in a field workshop) the system identifies the defective sub-components, replaceable in such workshop, or indicates the need of a higher level workshop. This level requires an off--board, portable diagnostic unit.

INTRODUCTION

Iveco has recently developed a complete power-pack for a tank application, as a part of a complete project in conjunction with Oto-Melara for the production of a new tank (named "Ariete") for the Italian Army. In this project also the study of suitable diagnostic systems was included; a power-pack diagnostic system has been then developed as well.

DESCRIPTION OF THE POWER-PACK

Even if a full description of the power-pack on which the diagnostic system has been applied is behiond the scope of this paper, a short introduction is necessary to understand the concepts of the diagnostic system itself.

A power-pack is a unit that includes almost all components of the power-train of a military vehicle, connected to the vehicle in such a way to allow easy replacement of the unit on the field.

In this case, the power-pack (named PPV120 - see fig.1) has been designed for a 50 t tank and includes:
- engine
- transmission and steering system
- cooling system for both
- auxiliary units

ENGINE - The engine used is a turbocharged diesel engine with a rated power of 1270 HP at 2300 RPM. Its main characteristics are:
- 12 cylinder, V-type engine, with a V angle of 90^ . Cylinder bore 145 mm, stroke 130 mm, for a total displacement of 25.7 l.
- two exaust gas turbochargers, one for each cylinder bank. Each turbocharger is followed by an air-to--water aftercooler.
- direct injection diesel cycle, with in-line high pressure injection pump mounted in the cylinder V.

TRANSMISSION - The transmission is a ZF type LGS 3000. It includes the following functions:

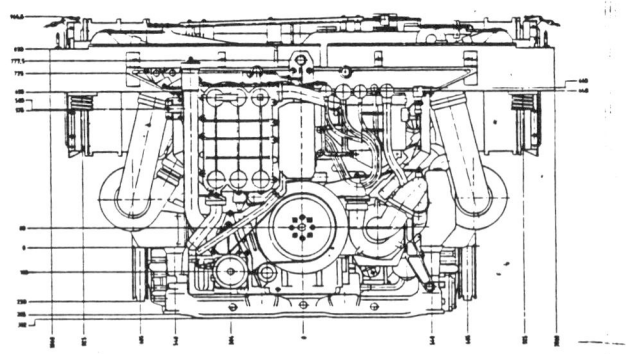

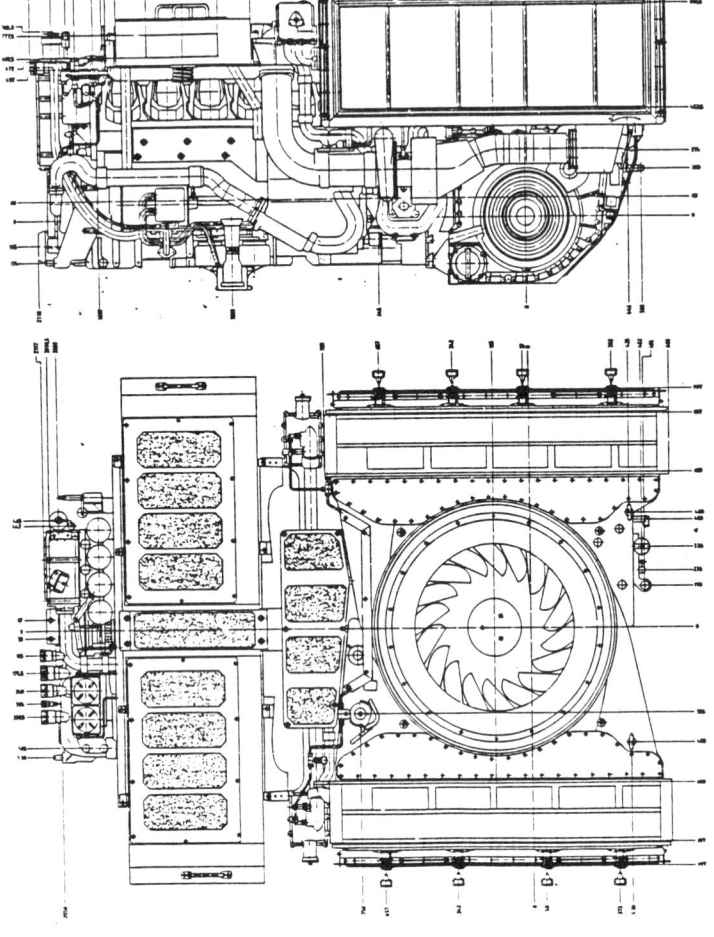

fig.1
Power-pack PPV120

- hydraulic torque converter with mechanical lock-up
- epicycloidal main gearbox with 4 forward and 2 reverse speed
- epicycloidal steering gearbox with 3-radius steering control on both sides and pivot capability (the tank is track-driven)
- hydraulic retarder
- 4 speed cooling fan drive

All functions are hydraulically controlled. The main transmission, torque converter, retarder and fan drive are activated with electrovalves. The steering transmission is directly activated with a mechanical link to the vehicle steering wheel.

The final reduction gears and the disk brakes are mounted on the tank body and are not included into the power-pack.

COOLING SYSTEM - An integrated cooling system is used for all heat sources on the power-pack; this includes the engine itself (cylinders and cylinder heads), the engine oil heat exchanger, the transmission oil heat exchanger (retarder and torque converter included), the two air-to--water aftercoolers, the air conditioning system heat exchanger.

Two large radiators, mounted above the transmission, are used to dissipate all the collected heat. A centrifugal fan, driven by the transmission, blows air through the radiators; a variable speed drive is used to minimize loss of power and noise, reducing fan speed when possible (the fan uses about 150 HP at full speed).

The system is completed by a water pump, driven by the engine, and two thermostats for quick heat-up of the system.

The complete cooling circuit is closed into the power-pack and is not opened or drained when the power-pack is removed from the tank.

AUXILIARY UNITS - The power-pack is complete of the following auxiliary units:
- two intake air filter
- a 12 kW 28 V electric generator
- two electric starter motor (both necessary to start the engine)
- some components of the air conditioning unit

CONTROL SYSTEM - The basic scheme of the electrical system of the tank, for the part regarding the power-pack, is shown in fig.2. The main units are
- the Power Control unit, which includes all electrical interfaces to the power-pack, for sensors, actuators and electrical power distribution (this unit controls also some other tank electrical sub-systems)
- the Dashboard unit, which includes most of the driver interfaces
- the Gearbox Control unit, which controls gear insertion on the gearbox

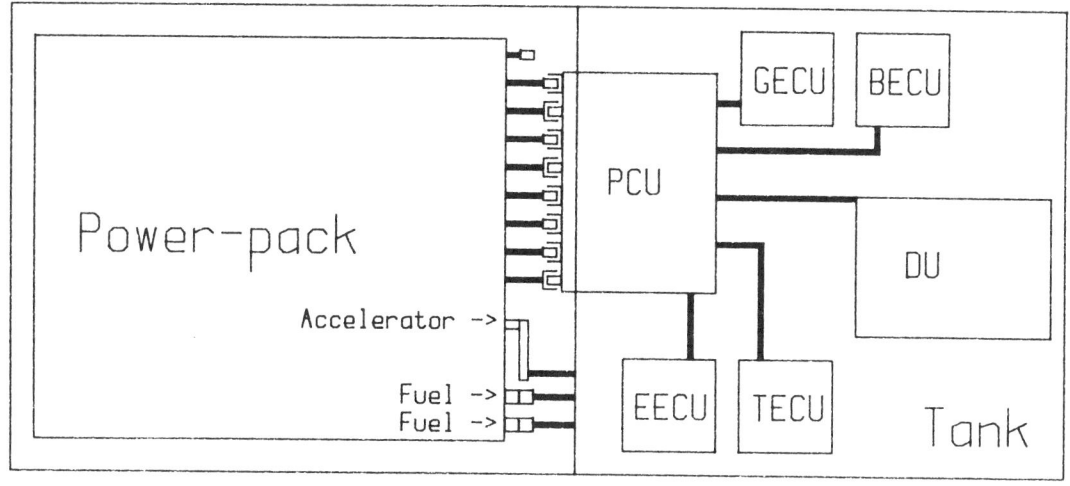

PCU = Power Control Unit
GECU = Generator Electronic Control Unit
BECU = Brake Electronic Control Unit
EECU = Engine Electronic Control Unit
TECU = Tranmsmission Electronic Control Unit
DU = Dashboard Unit

fig.2
On-board tank system

- the Generator Control unit, which controls generator voltage and current
- the Engine Control unit, which controls engine power reduction (de--rating) at high coolant temperature and which controls cooling fan speed
- the Brake Control unit, which controls the split of braking power between disk brakes and hydraulic retarder

Most of the units are microprocessor controlled, allowing application of sofisticated control strategies at all levels.

CONNECTIONS WITH THE VEHICLE – Excluding the main mechanical coupling (support and sprocket output shafts), the connections between power-pack and tank are localized on its front end, and are the following:
- 7 electrical multiple connectors for all electrical systems (power, actuators and sensors).
- 2 hydraulic coupling for fuel supply and return.
- 2 hydraulic coupling for the air conditioning system.
- a mechanical link for accelerator control.
- a mechanical link for steering control.

PRINCIPLES OF THE DIAGNOSTIC SYSTEM

A diagnostic system for a military, armoured vehicle cannot be developed and analized as an isolated system, but is something that is deeply correlated on one side to the complete vehicle development and on the other side to the military logistic organization.

It is well known that for this type of vehicles a good reliability is essential; the two main objectives to reach in this field are :
- to reduce possibility of vehicle failure during its use
- to increase vehicle availability by reducing repair time

These objectives are predominant over economical considerations, and are the main reason for the differencies between "military" and "civil" technology.

Of course these objectives must be in first place reached with a high "development" and "production" quality of the vehicle: the target is always a vehicle with a minimum number of failures and repairs in its useful life.

As the "perfect" reliablity cannot be reached, some help comes from the "preventive maintenance" concept:

it is preferable to replace a component with a defined life before its failure than to wait for the final failure. This is done to a limited extent in the civil field, as it will in any case lead to replace something which could still have some usefull life; it is much more applicable in the military field, where a "planned" repair is in any case preferable to an unexpected failure, even if it increase operational cost.

The task of the diagnostic system starts when all these actions fail; based on the previous objectives, the tasks of the diagnostic system become:
- to identify potential causes of vehicle failure before they become evident to the driver or they cause performance degradation: in any case it is preferable to know before a mission that a vehicle has a potential problem (and eventually replace that vehicle), than to have a failure during the mission itself.
- to reduce the time of vehicle non-availability for repair work: this means:
  * to identify the level of the fault (not all faults can be repaired in all places)
  * to identify the defective component
  * to verify the funcionality of the system after the repair

This is particularly important but difficult at the first level of intervention, on the field, where time is critical, equipment must be limited and expert personal is not available.

For this last reason this diagnostic system has been mainly developed for the first two level of intervention: it is then clear that the specification of the system is based on the following steps:
- define the "environment" of the level of intervention
- define the repair possibility at each level
- define the capability of the diagnostic system

As already mentioned, this system is dedicated to the vehicle power-pack, and does not cover all the other vehicle systems.

DESCRIPTION OF "LEVELS" OF REPAIR - The logistic organization of the Army has some differences from country to country; the present work has been based on the Italian Army structure , but in any case it could be applied with small changes to any Army.

The technical assistance to a tank is developed at various levels, with increasing capabilities but reduced mobility.

The "first level" units are acting directly on the field of operation: they are composed of few people (2-3), with a limited number of tools and a small dotation of spare parts; their task is to perform small repair and maintenance job directly on the tank on the field. As a reference such jobs should not take more than 60'.

The "second level" units are small mobile workshops with a more comprehensive dotation of tools and spare parts. These are still field units, which operates directly behind the field of operation, without any fixed installation.

All higher level units are fixed units, often specialized in specific fields of activity (i.e. transmissions), which operate far from the field. Here units returned as defective form the lower levels are repaired.

POSSIBLE OPERATIONS AT DIFFERENT LEVELS - Based on the above descriptions, some limitations on the extent of operations at the various levels can be easily defined.

First level - At first level the most severe limitation is the lack of accessibility to most components of the power-pack when it is installed into the tank. The only possible operations on the power-pack, without removing it from the tank, are basically:
- replacement of filters for engine oil, intake air and fuel
- re-filling of engine and transmission oil, cooling fluid

Furthermore some components which are related to power-pack functions but which are installed on the tank can be replaced: these are:
- electronic control unit for transmission and generator
- fuel supply pumps
- transmission control selector

For all remaining operation the power-pack must be removed from the tank; this is still considered a "first level" operation, possible on the field, even if it requires an external support like a "service" tank, equipped with a crane suitable to lift the power-pack itself, which weights over 5 t.

It is evident that the fastest way (even if not the more economical one) of repair after power-pack removal, is the replacement with a new, running unit without any attempt of further first level repair. The "service" tank can be equipped with a spare power-pack for such circumstances, allowing fast recovery into service of a defective tank (power-pack replacement will take less then 1 hour with adequate lifting equipment).

Second level - With the above conception of first level operation, second level (field workshop) units should deal only with isolated power-packs.

Even if second level equipment is more complete, due to the operation in field, some limitations are necessary to field repair, to avoid possible severe damages. These are basically:
- internal engine and transmission components must not be exposed to possible contamination (dust, water, etc.)
- critical bolts must not be loosened (engine heads, crankshaft bearing caps, flywheel, etc.)

Even if these are quite stringent limitation, a great deal of sub-components can still be replaced in field; these are:
- cooling system components (heat exchangers, pump, thermostats, cooling fan, pipes and seals).
- injection system components (injection pump, injector, pipes)
- turbocharging system components (turbochargers, manifolds, aftercoolers)
- electrical components (generator, starter motors, sensors, cabling, etc.)
- electronic engine control unit (this unit is physically replaceable also at first level, but it requires a calibration with the injection pump at which it is coupled, possible only with special equipment)

Furthermore at this level it is possible to dismantle the power-pack into its two basic sub-system (engine and transmission) and to replace one of the two if necessary.

DIAGNOSTIC BASIC SYSTEM SPECIFICATION - From the above description the specification for the complete diagnostic system can be worked out.

The "hardware" specification for the first level system is defined by two basic considerations:
- on the field auxiliary equipment must be reduced as far as possible
- the on-board electronic system of the tank is sophisticated and flexible.

The conclusion is quite logical: the more efficent way to introduce a first level diagnostic system is to have it incorporated into the basic tank on-board system, in such a way to be operated without any external equipment.

For the second level, a different consideration becomes important: at this level the test will be basically performed on a power-pack dismounted from a tank, and the tank will not be available; as it is clear from the power-pack description, to run test under such circumstances is feasible (the power-pack is a "closed" unit and the engine can be operated) but some auxiliary units are necessary (a mechanical support, batteries for engine starting, a fuel tank). The second level system must be able by itself to control all power-pack functions in such an "insulated" test.

DESCRIPTION OF THE DIAGNOSTIC SYSTEM

ON-BOARD SYSTEM - As the diagnostic system for the first level is installed and integrated into the tank on-board system, its description requires some analysis of the complete on-board system.

From a "hardware" point of wiew the on-board system has already been described. Some more details are however necessary to describe the available interface for information input/output (Dashboard Unit, fig.2). The tank dashbord is fully digital, and includes:
- some indicators, both fully digital and of the "bar" type
- a set of indicating light
- a small alpha-numeric display
- some input switches

This dashboard is utilized for all on-board functions, as will be described in the following.

Continuous information - At any time electrical power is available, the dashbord gives following information:
- Engine speed, as digital value and "bar" display
- Vehicle speed, as above
- Coolant temperature, engine oil

temperature, engine oil pressure, as "bar" display
plus some additional "vehicle" information (fuel level, indication of status of various sub-system, etc.).

Alarm conditions - Here the principle of giving an alarm indication during normal vehicle use only if the corresponding fault will degrade vehicle performance has been applied; for example, a clogged filter requires maintenance but does not alter vehicle operability and thus is not shown as alarm when in driving conditions.

The possible alarm conditions, related to power-pack operation, are:
- low engine oil pressure, low gearbox oil pressure
- low coolant level (two levels of alarm)
- high engine oil temperature, gearbox oil temperature, coolant temperature (the later with two levels of alarm)
- control unit fault for transmission, brakes, engine, generator.

The conditions of low engine oil pressure and low coolant level (level 2) cause as well automatic engine stop. A special switch ("battle switch") can exclude such protection.

Some special consideration is necessary for the conditions of over-temperature, as these can be intermittent and in any case will disappear when the engine is stopped and cooled. On the other side a condition of over-heating is an indication of a fault and must be corrected before re-use of the vehicle. To avoid problems connected to a "lost" over-temperature warning, such alarms are permanently stored in non-volatile memories and will be shown at every following diagnostic test. For this functions, as all the other "permanent" recording of data, the memories are located into the Engine Control Unit, which in any case cannot be separated by the power-pack for calibration reasons.

The same storage is not done for all other alarms, which are in themselves non-volatile (i.e. a low level will never go back alone to "high").

Start-up test - At every system power-up, a certain number of functions are automatically checked; these are essentially sensors conditions and electronic control unit conditions.

For sensors, the type of test depends on the sensor type and characteristics: the base principles are the following:
- for "active" sensors, i.e. sensors with built-in electronics, sensor calibration is such to have a pre-defined output offset at "zero" input (engine not running). If the offset is correct, cabling and power supply integrity is guaranteed. Sensor integrity is verified for the majority of possible failure modes. This class of sensors include all pressure sensors and some speed sensors.
- for temperature sensors, in most cases a double sensor, with two sensing elements in one housing has been employed. A check of coherence between the two elements guarantees cabling and sensor integrity. For exhaust temperature, where thermocouples are employed, the coherence between sensors on the two cylinder banks is verified.
- for most on-off sensors a full test is impossible. Where output status at engine not running is defined, this is checked (i.e. oil pressure is low with engine stopped); depending on type of cabling, this could detect short circuit but not open circuit or vice-versa. In some cases coherence between different sensors can be used as a further check.

For electronic control units, a self-test is performed at power-up and an OK signal is given to the tank system if the control unit is correctly functioning.

In all cases the result of the power-up test is not affecting the possibility for starting the engine; in other words the display on the dashbord will indicate the existence of a fault, but the driver can decide to start the engine and to use the vehicle also in such cases. The same applies for all following diagnostic tests.

GO-NOGO test - After system power-up the driver can ask for a more comprehensive test of power-pack conditions. This is done by pressing the "TEST" button on the tank dashbord. This test is composed by a sequence of tests, automatically performed by the tank on-board system (see fig.3).

As the tank on-board system has no possibility of directly actuating some control (accelerator, brakes, steering, gear engagement) it requires specific driver actions when necessary via the alpha-numeric display.

The test sequence starts with some preliminary tests, before allowing any attempt to start the engine. These are:
- test of steering and acceleration position sensor. This is done by requiring to the driver certain actions and verifying the sensor response.
- check of minimum gearbox and engine oil level, to allow safe power-pack starting.
- check of fuel supply pressure and fuel filter clogging (fuel is supplied by two electrical pumps also before engine starting).
- check of memorized over-temperature warnings.

If the results are correct, a "controlled" engine start is performed; the step are the following:
- first the engine stop magnet is engaged and its correct functioning verified
- without releasing the stop magnet, the driver is required to activate the starting motors and to crank the engine.
- the system checks engine speed and battery voltage; if speed and voltage are correct, the stop is released.
- after a certain time, correct engine starting is verified.

If the engine does not start, this procedure allows the identification of the failure. In such a case, the procedure is repeated three times.

In case of correct starting, a first pre-heating phase is initiated, by asking to the driver to activate vehicle brakes, engage top gear and accelerate the engine to 1100 RPM. In this way the transmission torque converter is used as a "brake", by absorbing all power developed by the engine. However the prescribed conditions are such that a limited power is dissipated, with no possibility of damage to the transmission. This condition is mantained up to 50^C water temperature.

When this temperature has been reached, the transmission is disengaged and the driver is requested to fully accelerate. In this condition the maximum engine speed, the engine oil pressure and oil filter clogging are checked (in any case, a low engine oil pressure warning is always active and will stop the engine in case of a serious fault in the lubri-

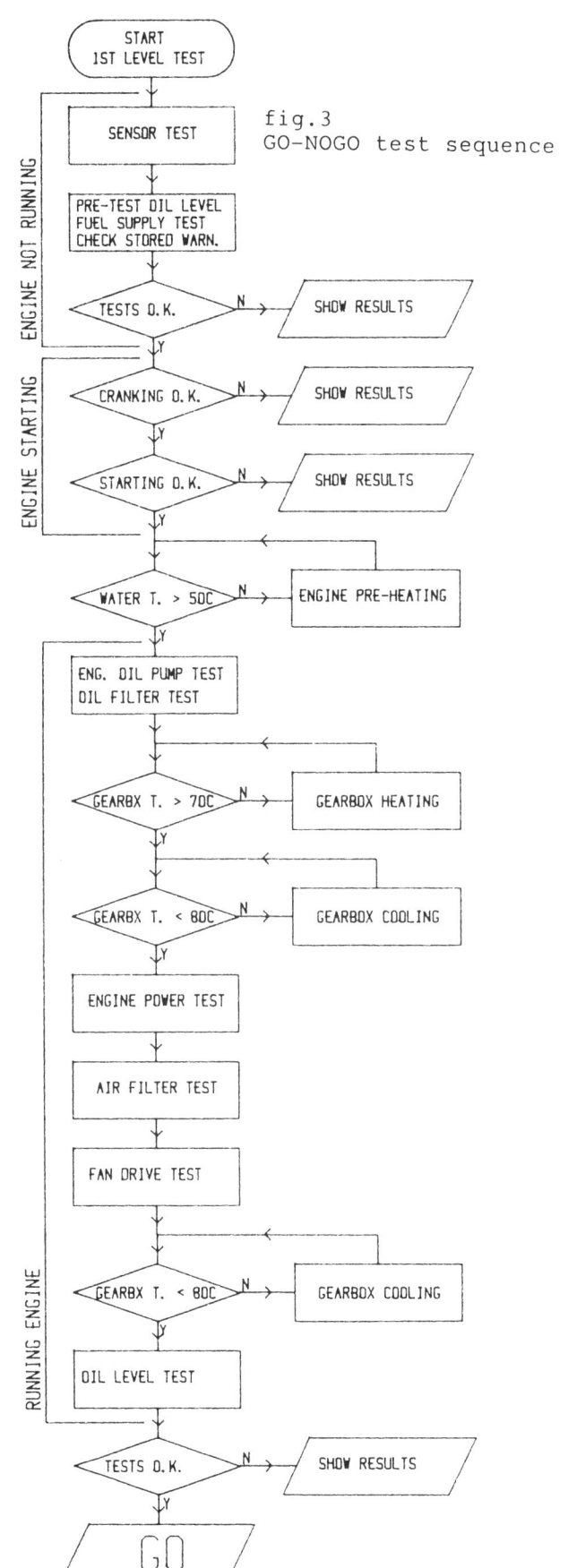

fig.3
GO-NOGO test sequence

cation system).

After this phase the engine power test is measured, by using the torque converter of the transmission as a dynamometer. All tests performed have shown that the relation torque/speed on the input shaft of the hydraulic torque converter with locked output is repeatable and accurate, provided that the oil temperature into the transmission is in a controlled range, This means that the speed reached by the engine while fully accelerating with a gear engaged and brakes locked is in good correlation with the full engine power. Furthermore, the time necessary to reach such a condition from idle gives additional indications on the performances of the turbocharging system.

In this on-board system a simplified procedure is used, as follows:
- if the transmission oil temperature is outside the prescribed interval of 70 to 80 ^C, a suitable heating or cooling cycle is performed. The heating cycle is similar to the one used in the previous phase (with a slightly higher engine speed to accelerate oil heating), while cooling is obtained running the engine at 1200 RPM no load, with the cooling fan at maximum speed.
- the driver is asked to disengage transmission and accelerate fully three times to pre-heat the exahust system (this improves repeteability of the test).
- the driver is asked to engage 4th gear, fully press the brake pedal and then fully accelerate.
- the test is considered satisfactory if in a given time the engine reaches a certain reference speed, indicative of good performances (indicatively, the "low power" threshold is - 5 % from original power). To increase accuracy the actual "pass level" of the power-pack under test is used as reference, as this value is permanently memorized into the Engine Control Unit.

The test is interrupted if a maximum test time is exceded (indicatively 18"), the maximum transmission temperature is exceded (140 ^C) or a vehicle movement is detected by the vehicle speed sensor.

At the end of the power test, before asking to the driver to release the accelerator, the air filter clogging is verified (under these conditions intake air flow is maximum).

Then the driver is requested to release accelerator and disengage gear; the cooling fan is activated at full speed and its speed checked to verify fan drive conditions. As a final test, engine and transmission oil level are checked under "hot" conditions.

The test is then completed with a "GO" message, or with the indication of proper repair actions (1st level diagnostic).

Some remarks are necessary:
- all "pass" levels are not set to a real "limit" value, but they are set to such a value to allow still the use of the vehicle for a typical "mission". For example, the "minimum engine oil level" is set with a margin to the absolute minimum corresponding to the typical oil consumption for 8 hours operation.
- all phases where driver actions are required, are performed with suitable "time-outs" to discontinue the test and avoid dangerous situations if the driver does not perform the required action. Furthermore, for the engine power test (the most dangerous for power-pack integrity in case of mis-use), the system can reduce engine power indipendently from the driver, by fully activating the power reduction unit.
- as already mentioned, even with a "NO GO" result of the test, the tank can be normally operated if this is required. The test can be discontinued at any time, by resetting system power.

1st level diagnostic - First level diagnostic information, based on the go/nogo test, is shown on the tank alpha-numeric display, at the end of the test. Possible indications are:
- low transmission or engine oil and coolant level (refill)
- battery charge too low to start engine (recharge/replace)
- clogged air, fuel or engine oil filters (replace)
- defective electronic control unit for generator or gearbox (replace)
- defective fuel supply pumps (replace)
- defective power-pack (replace).

The conditions which lead to a "replace power-pack" (i.e. second level fault) message are:
- defective sensors / cabling
- defective engine control unit

- over-heating during previous use of the tank
- defective generator, starter motors, stop magnet
- fault in lubrication system
- clogged transmission oil filter (not accessible with power-pack into the tank)
- defective fan drive
- defective gearbox
- low engine power

Even if not strictly necessary, at this level some indication of the second level fault is given, when possible. For example, in case of a defective sensor, the sensor is identified even if the replacement will be possible only at second level.

OFF-BOARD SYSTEM - SECOND LEVEL - As already defined, the second level diagnostic test must be executed with an external unit, which allows to run the power-pack indipendently from a tank, while performing all necessary measurements.

To run a power-pack in this conditions, as can be understood from the power-pack description, the following equipment is necessary:
- a mechanical support for the power-pack
- a fuel tank with an electrical fuel supply pump
- a 24 V battery pack for engine starting
- an electrical control unit.

During the development phase, it has been found quite logical to incorporate all control and measurement functions in one single unit, in such a way to have an integrated automatic control of the test.

This unit has been specially developped for this application, and its description can be of some interest.

Diagnostic unit - This unit is mechanically divided into three major sub-units; in this way it is fully portable and of easy use on field. These units are:
- the Interface Unit (IU), which includes all power cabling, high current relays, current sensors and specific sensor interfaces. This unit is coupled to the power-pack with the 7 connectors normally used for the connection to the tank, plus an additional connector for some sensors which are specific for the 2nd level and are not connected into the tank. Externally to the unit, a generator control unit is installed, to allow proper generator operation during tests.
- the Data Acquisition Unit (DAU), which includes all analogue and digital input/output circuits, including signal conditioning, insulation amplifiers, multiplexers, D/A and A/D converters, and a complete microcomputer dedicated to the on--line test control and data acquisition. This unit is connected to the previous by five cables.
- the Control Unit (CU), which includes the operator inteface (an LCD screen plus keyboard) and a microcomputer dedicated to test sequence programming and data analysis. For data analysis an "expert system" is included into the software of this unit.

All control functions are performed via the Control Unit, and the other two units do not have any external control, with the exception of an emergency stop button on the Interface Unit.

The system is completed by an Accelerator Actuator Unit (AAU), which is mechanically coupled to the accelerator linkage on the power-pack. In this way the system has a complete control over all power-pack functions.

The division into three sub-unit of the system gives, in addition to better portability, a second advantage: only the Interface Unit includes hardware which is specific for this power-pack (connectors layout, special sensors, etc.), while the Data Acquisition Unit and the Control Unit can be considered an "intelligent", generic data acquisition system, which could be (in future) used for different types of power-pack. For this scope the number of I/O channels on the DAU has been fixed with some over-capacity, in view of future developments.

Electrical supply specifications for the system are the same of a 24 V on-board system; during the test electrical power can be supplied by the same battery pack used for engine starting and by the power-pack generator, when running. However, to allow testing of power-packs with serious electrical malfunctions (i.e. starter motor short circuit), an additional 24 V supply can be connected for system operation.

All components of the system are of course designed in accordance with military standards for outdoor field operation (vibration, temperature,

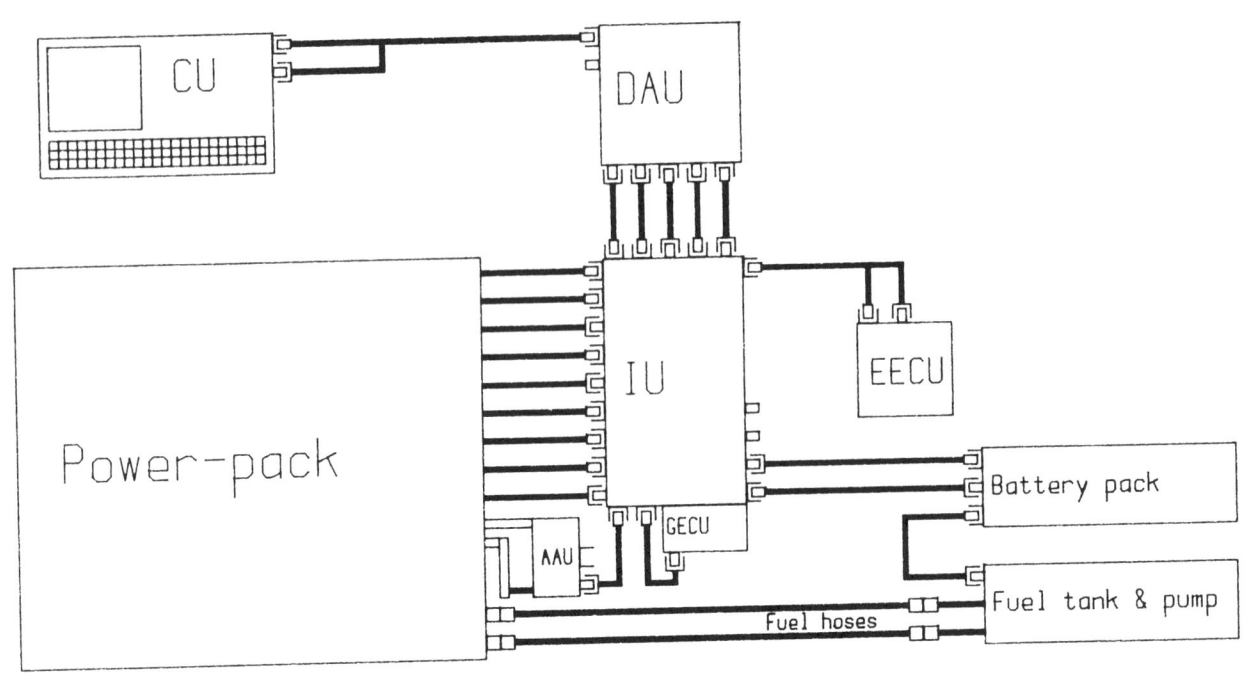

IU = Interface Unit
DAU = Data Acquisition Unit
CU = Control Unit
AAU = Accelerator Actuator Unit

DIAGNOSTIC UNIT

EECU = Engine Electronic Control Unit
GECU = Generator Electronic Control Unit

fig.4
Second level unit
Basic configuration

humidity, etc...). The Control Unit, which includes a screen and keybord, is completed by additional covers for protection during storage and transport.

Possible system configuration - The basic design configuration of the unit is shown in fig.4: the test is performed indipendently from a tank, the diagnostic system provides all control and monitoring functions, fuel and electrical power are supplied by external units, the power pack is mounted on its support.

However the system has been designed to provide as well some additional configuration, which could be of some help in the practical use of the system.

The first possibility (fig.5) is to replace the auxiliary fuel tank and battery pack with the tank to which the power-pack belongs, with the power-pack outside the tank on its support. In this way the test equipment can be installed in a faster way, especially in adverse environments (battery pack and fuel tank are heavy units, while the diagnostic system in itself is portable), but the tank cannot be returned to service (with a new power-pack) as long as the repair is not completed.

The second possibility (fig.6) is to perform the test with the power-pack into the tank, without any external support except the diagnostic unit. Even if this option seems quite attractive (no heavy equipment is necessary), it has some serious drawbacks :
- if a second level test is performed, a fault is in any case present, which will require power-pack removal; the gain in time to perform the test into the tank is then only apparent.
- the connection of the system requires in any case the removal of the power-pack compartment cover; this is already a significant percentage of the total time required to replace the power-pack.
- the electrical connections are located in a very narrow space, and the connection of the system will be a difficult task.

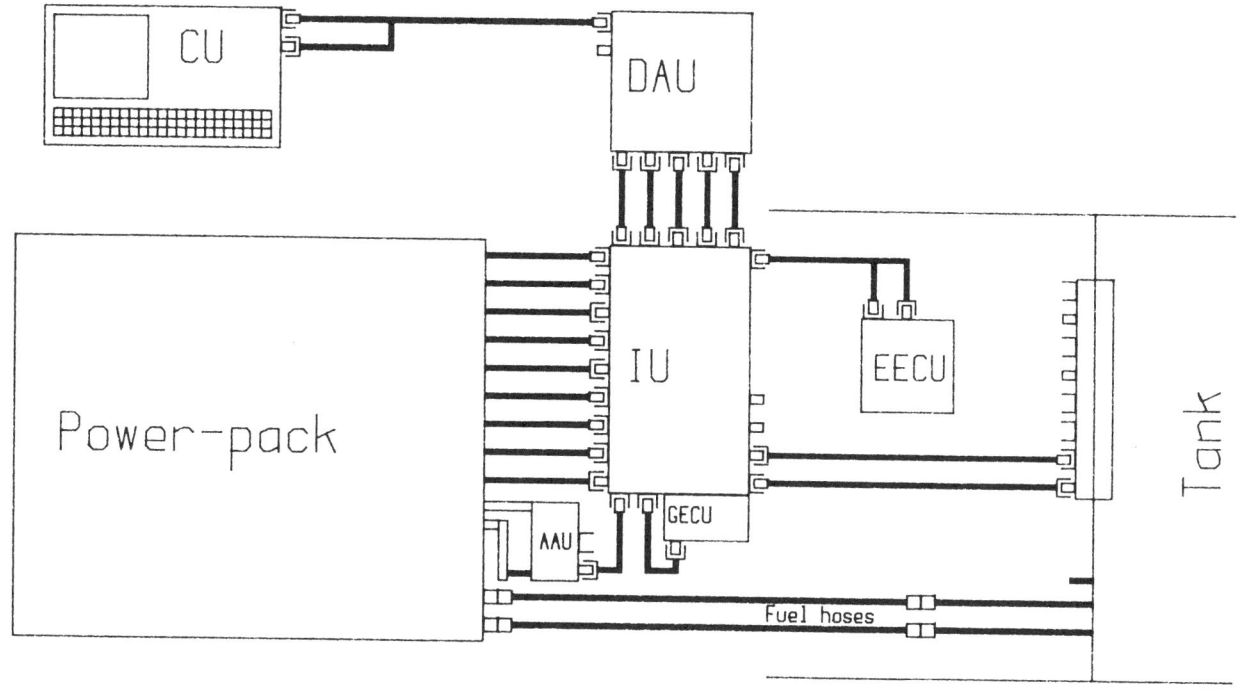

fig.5
Second level unit
1st variant

As in any case this possibility requires only a few additional cables, it has been considered and it will be practically tested.

It is important to notice that, even when the tank is used as a support for second level testing, all control functions will be performed by the diagnostic system, the tank beeing only a source of fuel and electrical power.

Functions of the diagnostic unit
- The diagnostic unit is capable of many functions, not only for diagnostic use, but also as a help during repair and maintenance work. In detail, the options considered up to now are:
- to peform the same tests that are performed at first level on the tank, in a fully automatized way
- to perform the second level test sequence, with more complete diagnostic capabilities
- to perform injection timing measurements
- to perform power reduction unit adjustment
- to reset stored warnings and to change parameters in the permanent memory of the Engine Control Unit

Of these options, the most relevant and of interest is the 2nd level diagnostic test, which will be described now.

Second level diagnostic test -
As already mentioned, the diagnostic unit is capable of fully controlling power-pack functions. This means that the test sequences are in this case fully automatic, and no operator action is required (fig.7).

As a first step, a sensor test is performed, with the same principles of the first level test. However, in this case also the sensors specific for second level are tested (this is impossible in the tank, as they are not connected to the tank on-board system). In case of a defective sensor, the test is interrupted as the test requires all sensors functioning.

As a following step, an engine starting procedure similar to the fi-

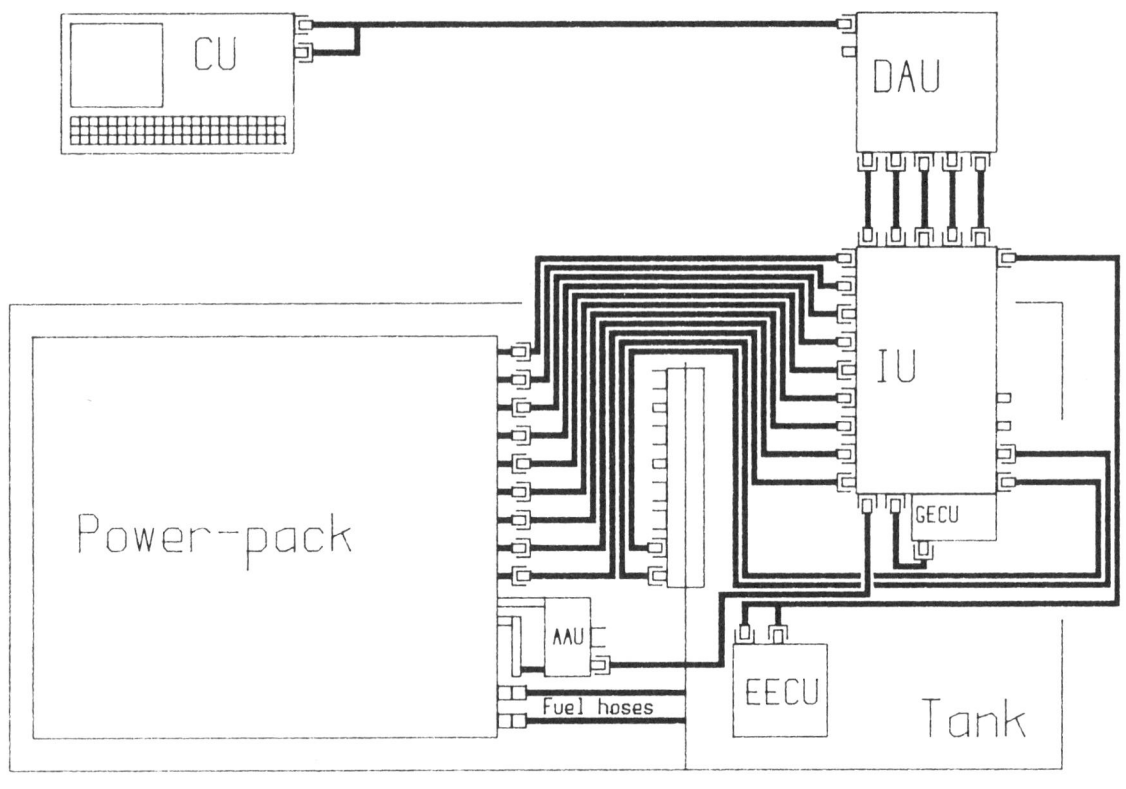

fig.6
Second level unit
2nd variant

rst level one is applied. However, some additional steps are added to improve fault detection:
- the cold start system (flame start) is tested; this requires ignition of the flame start system before cranking the engine and measurement of inlet manifold air temperature during cranking. Furthermore, the current flowing into the flame start heat plugs is measured.
- the current flowing into the starter motors and into the starter motors solenoids is measured, to identify accurately any fault into the starter system.
- during cranking, before starting the engine, an engine "compression" test is performed. This requires measurement of engine istantaneous speed during two engine revolutions. In this situation (engine not firing) speed fluctuations are correlated to cylinder compression ratio, and a low compression ratio on one cylinder can be easily identified. As a matter of fact, due to the structure of this engine (12 cylinder with a V angle of 90^) that cause uneven spacing of combustions, only a defective cylinder couple can be identified. Faulty cilinder location requires some further test.

If the engine does not start, of course the test is terminated. However all collected data will allow identification of the reason for the failure, at level of defective component. If the engine starts correctly, the cooling system test is performed.

The cooling system test is divided into two phases. In the first phase, an engine heating cycle similar to the one at first level is performed (torque converter acting as a dynamometer, engine at low speed and low load); as in these test the power-pack is not into the tank, the vehicle disk brakes cannot be used to lock gearbox output; the locking is then obtained by engaging simultaneously two different gears (this is possible as the diagnostic unit has full control on the gearbox control valves as well). The water temperature is increased up to 80 C^, and it is continuously measured; the change in slope corresponding to thermostat opening is detected and recorded. This test is skipped if the initial water temperature was above 50 ^C, as this will not allow detection of the opening point. In this case the engine is simply heated up to 80 ^C.

After this, the engine is unloaded and the fan inserted to full speed; the slope of the following cooling curve is measured as an indication of overall cooling system efficency.

As following phase, the Engine Control Unit is verified in all its functions. This test has the following phases:
- check of derating function : different water temperature are electrically simulated, and the corresponding movement of the power reduction actuator measured.
- check of fan control function : different water temperature and control inputs are simulated, and the corresponding outputs to fan drive control valves measured.
- check of power reduction actuator : a water temperature corresponding to full derating is simulated, and a power test, silmilar to first level test is performed, to verify actual power limitation.

The overall engine friction is then measured: the engine is accelerated to full speed, and then fuel is cut out; from the law of decelleration of the engine, knowing its inertia, the friction torque can be evaluated.

Then a full power test is performed. The tecnique is always the same (locked torque converter used as dynamometer, fully accelerating the engine), but a number of parameters are measured during the test; these include:
- speed, boost pressure for both cylinder banks, fuel rack position during the complete acceleration phase
- exhaust gas temperature (a value for each group of three cylinders, as allowed by the exhaust mainfolds design), inlet air temperature, pressure drop through air filters, etc... at full power
- istantaneus speed fluctuations over two engine revolutions at full power to identify a faulty cylinder

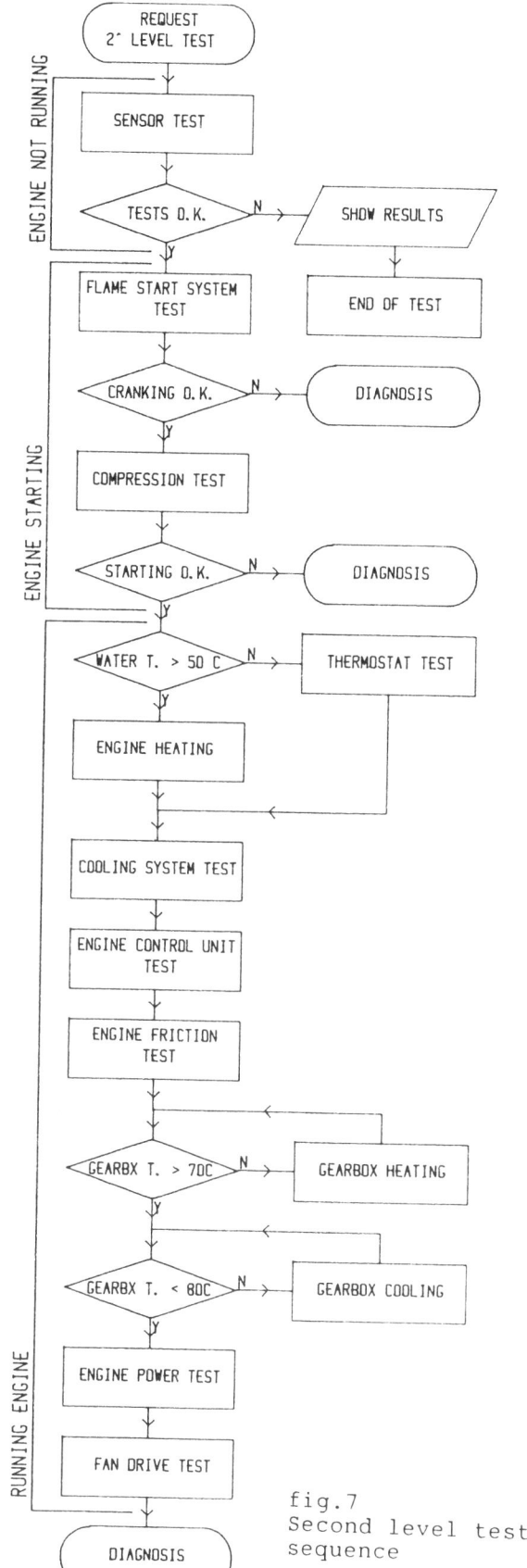

fig.7
Second level test sequence

(the tecnique is similar to the one described for the compression measurement - again only cylinder pair can be identified)

As for first level test, a certain number of "safety" parameter are monitored and the test is interrupted if some dangerous situation is detected (high transmission oil temperature, movement of gearbox output, etc...).

As a final test, the cooling fan drive is tested; in this case the engine is run at full speed and the various speeds are engaged. For each speed the actual fan speed is measured.

Diagnostic procedure - For this second level test, the fault identification is not in direct correlation with the result of the measurement; some faults can be only identified by correlating results of different tests; for example, a cylinder with a significantly lower power will be identified by cross-correlating the result of istantaneous speed measurement (which identifies a couple of cylinder) with a low exhaust gas temperature (which identifies three possible faulty cylinders); then the correlation with the compression test will indicate if the cause is a low compression ratio on that cylinder.

To allow easy implementation of such procedures, an "expert system" has been incorporated into the diagnostic unit. This system analyzes all the data acquired during the test phase, and based on a certain number of "rules" gives the indication of the faulty component and of the action necessary to recover the full funcionality of the system. Of course, in same cases the identification cannot be complete, and in such cases the system can give a series of possible causes, with a probability of occurrence.

The use of the "expert system" will also allow easy upgrade of the system, with the results of practical experience; its structure will allow simple introductions of new or different "rules" for fault identification and the possibility of giving higher "probability" to more common faults.

CONCLUSIONS

The system described in this paper has already been tested and demonstrated in laboratory with laboratory test equipment; all components are now in the industrialization phase, and some prototype tanks are alredy equipped with the first level system.

AKNOWLEDGEMENT

The work reported into the present paper is the result of a number of activities performed inside different groups of our Company, as well from external Companies. In particular, we thank:
- Italian Army Technical Staff for the tecnical support
- Oto-Melara for the implementation of the system on the tank on-board units
- SEPA for the development of the off-board unit

Positions and opinions advanced in this paper are those of the author(s) and not necessarily those of SAE. The author is solely responsible for the content of the paper. A process is available by which discussions will be printed with the paper if it is published in SAE Transactions. For permission to publish this paper in full or in part, contact the SAE Publications Division.

Persons wishing to submit papers to be considered for presentation or publication through SAE should send the manuscript or a 300 word abstract of a proposed manuscript to: Secretary, Engineering Activity Board, SAE.

Printed in U.S.A.